NEUROSCIENCE RESEARCH PROGRESS

EVOKED POTENTIALS AND ELECTRICAL STIMULATION

CLINICAL ROLES, CHALLENGES AND EMERGING RESEARCH

Neuroscience Research Progress

Additional books in this series can be found on Nova's website under the Series tab.

Additional e-books in this series can be found on Nova's website under the eBooks tab.

NEUROSCIENCE RESEARCH PROGRESS

EVOKED POTENTIALS AND ELECTRICAL STIMULATION

CLINICAL ROLES, CHALLENGES AND EMERGING RESEARCH

TIMOTHY EISENHOWER

EDITOR

nova
biomedical
New York

Library of Congress Cataloging-in-Publication Data

ISBN: 978-1-53611-075-3

Published by Nova Science Publishers, Inc. † New York

CONTENTS

PREFACE

This new book discusses advances in research on evoked potentials and electrical stimulation. Chapters One describes the neurophysiological immediate and short-term changes that occur in the nervous system structures of patients assessed in three different contexts of work: Intraoperative Neurophysiologic Monitoring (IONM), ICU, and in the Clinical Neurophysiological Laboratory. Chapter Two focuses on the importance of cortical evoked potentials to analyze the neurophysiological correlates of migraines, strokes and related psychiatric complications, paying particular attention to the possible implications in terms of clinical applications and future research. Chapter Three analyzes electric stimulation in stroke patients. Chapter Four describes what is known about the effects of the use of electrical stimulation of carcasses of meat animals, including the effects on meat tenderness and meat sensorial characteristics. Chapter Five explores the effects of oxidative stress on brain electrical activity and its repercussions on sensory organization in geriatric rhesus monkeys in captivity. The final chapter provides a short communication on visually evoked potentials in rhesus monkeys.

Chapter 1 - Introduction. From findings in the neurophysiological tests, especially with the recordings of the Evoked Potentials (EPs) it is possible to characterize the functional state of the Nervous System (NS) structures, in the course of diseases, in the Clinical Neurophysiologicaly Laboratory, and also can be performed during surgery and at Intensive Care Unit (ICU).

Objective. To describe neurophysiological immediate and short-term changes that occur, in the NS structures in patients assessed in three different contexts of work: Intraoperative Neurophysiologic Monitoring (IONM), ICU, and in the Clinical Neurophysiological Laboratory.

Methods. Here are presented the results of the neurophysiological evaluation in a case-study of patients during their admission at ICU, IONM. In addition, the authors provide evidences of the value of EPs in assessing neuroplasticity in deaf and deaf-blind subjects. In each section Auditory Brainstem Responses (ABR), Visual Evoked Potentials (VEP) and Somatosensory Evoked Potentials (SSEP) were used according to their pathology. In addition, other neurophysiological tests such as Electromyography (EMG), Nerve Conduction Study (NCS), and Motor Evoked Potentials (MEPs) were performed according to the clinical situation of patients.

Results. It was possible to illustrate the practical utility of EPs in IONM and ICU. The neurophysiological changes are summarized using graphics showing the main findings obtained with EPs and other EDX tests. Electrophysiological responses change over time, being possible to provide evidences of the disease course through the measured parameters of each electrophysiological signal. On the other hand, the expansion of the SSEP cortical response to the left temporal region in subjects with sensory loss was interpreted as evidence of Cross-Modal-Plasticity.

Conclusions. EPs associated with other neurophysiological tests provide means to noninvasively brain functional assessment in patients at ICU and IONM. EPs are largely used for supplement information obtained from clinics, neuroimage methods, and other laboratory tests. Nowadays, SSEP and ABR are the most important EPs used, and show interesting findings in the patient's assessment at IONM and ICU, while a characterization of the topographic distribution pattern of peaks in cortical evoked responses yields a complementary value to EPs assessment and interpretation for clinic and research.

Chapter 2 - In the last years, cortical-evoked potentials have been extensively used to study any abnormal cortical information processing associated with a specific neurotransmission pathway (i.e., serotonergic neutrasmission), which could represent a key element of several neurological conditions.

In this view, the intensity dependence of auditory-evoked potentials (IDAP), and the habituation deficit of visual and auditory-evoked potentials (VEP and AEP habituation deficit, respectively), represent the two main cortical evoked paradigms adopted in neuropsychiatric research fields.

IDAP is thought to be inversely proportional to the central serotonergic tone; thus, it represents a useful non-invasive indicator of serotonergic activity associated with specific clinical and behavioural correlates. For this reason, many studies investigated central serotonergic tone alteration in several neuropsychiatric disease.

On the other hand, the habituation phenomenon of neural evoked activity represents a physiological decrement in the behavioural response that results from repeated stimulation and that does not involve sensory adaptation or fatigue. A deficit of habituation (i.e., reduction or absence of the normal decrease in cortical response after repeated sensorial stimulation) is thought to be a core feature of several neurological conditions, in particular of migraine, where lack of habituation (or even an amplitude increase in the cortical response) has been observed in the interictal phase with various stimulus modalities.

In the present chapter, the authors will focus on the importance of cortical-evoked potentials to analyse the neurophysiological correlates of migraine, stroke and related psychiatric complications as well, paying particular attention to the possible implications in terms of clinical applications and future researches.

Chapter 3 - Surface electrical stimulation is one of the tools suggested for the rehabilitation treatment of patients following a stroke. Although its use is not very widespread in current clinical practice, it is regarded as a non-invasive, low-cost, reliable and easy-to-use treatment which could stimulate recovery in stroke patients. In recent years, the understanding of motor learning, neuronal plasticity and functional recovery after brain damage has improved, and the new findings in neuroscience have given a fresh impetus to research in neurorehabilitation. The number of existing studies regarding electrical stimulation and the stroke patient shows the interest which the subject still provokes and the need to provide more evidence that will assure its beneficial effect in clinical practice. Current lines of investigation are principally focussed on the recovery of the upper-limb, lower-limb and gait, hemiplegic shoulder, dysphagia, urinary incontinence and visuo-spatial neglect. Additionally, different modalities of application are presented and the most effective protocols are investigated, including their complementary use with other therapeutic strategies and new technologies. In this chapter, these points are extensively discussed to showcase the latest evidence, shed more light on this intervention in the field of neurological rehabilitation and establish future lines of interest in which advances in neuroscience and technological development play a significant role.

Chapter 4 - In the last decades, nonhuman primates (NHP) have been widely used as biomodels to study cerebral aging in relation to oxidative stress (OS) and the way in which organisms respond to it, since it is known that during aging reactive oxygen and nitrogen species (ROS and RNS) increase and, at the same time, the activity of the antioxidant systems decreases. In this sense, it has been described that when the neuronal plasma membrane components are oxidized, important neurophysiological changes occur, which potentially alter a great variety of electrophysiological parameters, generating a decrease in the nerve conduction velocity. During the last decades, it has been shown that with age increment, different mechanisms are generated in the central nervous system (CNS) by which oligodendrocyte (OLG) cell damage and death occurs. OLG produce myelin sheaths that surround neural axons, favoring nerve impulse propagation. Myelin sheaths coating greatly increases axonal conduction velocity, so its loss or degeneration causes neuronal axons demyelination, which is associated to sensory pathways nerve conduction velocity delays. It is for this reason that the sensorial alterations that occur during old age are especially important, not only because of their high prevalence, but also because of the direct repercussions that are generated on living organisms. It is therefore of utmost importance to evaluate the relationship between age and oxidative increase and its effect on different neuronal generators of sensory pathways in a primate model.

Short Communication - Visual evoked potentials (VEPs) are a non-invasive technique used in the evaluation of the retino-geniculate-cortical response; such responses are derivative from the correct function of the structures belonging to the visual pathway, which are the retina, lateral geniculate body and visual cortex. VEPs, as being a current sensitive-based technique are also a suitable means for understanding pathologies developed from the injury or malformations throughout the pathway.

In: Evoked Potentials and Electrical Stimulation ISBN: 978-1-53611-075-3
Editor: Timothy Eisenhower

Chapter 1

EVOKED POTENTIALS: CHALLENGES FOR CLINICS AND RESEARCH

Lidia E. Charroó-Ruíz**, MD,
Alfredo Álvarez Amador, PhD, Mario Estévez Báez, PhD, Fernando Rivero Martínez, MD, Ana Calzada Reyes, PhD, Liana Portela Hernández, MD, Valia Rodríguez Rodríguez, PhD, Adonis Valero Sánchez, Sandra Bermejo Guerra, MD, Antonio Paz Cordovés, MD, Beatriz Bermejo Guerra, MD, Manuel Sevila Salas, MD, Elina Ceballos Villalón, MD, Yesy Martín García, Gerda González Sánchez, Idania Valdés Amalde and Lídice Galán García, PhD

ABSTRACT

Introduction. From findings in the neurophysiological tests, especially with the recordings of the Evoked Potentials (EPs) it is possible to characterize the functional state of the Nervous System (NS) structures,

* Corresponding author: Lidia E. Charroó-Ruíz. Clinical Neurophysiologist. Associate professor, Cuban Center for Neuroscience, Havana, Cuba, lidia.charroo@infomed.sld.cu.

in the course of diseases, in the Clinical Neurophysiologicaly Laboratory, and also can be performed during surgery and at Intensive Care Unit (ICU).

Objective. To describe neurophysiological immediate and short-term changes that occur, in the NS structures in patients assessed in three different contexts of work: Intraoperative Neurophysiologic Monitoring (IONM), ICU, and in the Clinical Neurophysiological Laboratory.

Methods. Here are presented the results of the neurophysiological evaluation in a case-study of patients during their admission at ICU, IONM. In addition, we provide evidences of the value of EPs in assessing neuroplasticity in deaf and deaf-blind subjects. In each section Auditory Brainstem Responses (ABR), Visual Evoked Potentials (VEP) and Somatosensory Evoked Potentials (SSEP) were used according to their pathology. In addition, other neurophysiological tests such as Electromyography (EMG), Nerve Conduction Study (NCS), and Motor Evoked Potentials (MEPs) were performed according to the clinical situation of patients.

Results. It was possible to illustrate the practical utility of EPs in IONM and ICU. The neurophysiological changes are summarized using graphics showing the main findings obtained with EPs and other EDX tests. Electrophysiological responses change over time, being possible to provide evidences of the disease course through the measured parameters of each electrophysiological signal. On the other hand, the expansion of the SSEP cortical response to the left temporal region in subjects with sensory loss was interpreted as evidence of Cross-Modal-Plasticity.

Conclusions. EPs associated with other neurophysiological tests provide means to noninvasively brain functional assessment in patients at ICU and IONM. EPs are largely used for supplement information obtained from clinics, neuroimage methods, and other laboratory tests. Nowadays, SSEP and ABR are the most important EPs used, and show interesting findings in the patient's assessment at IONM and ICU, while a characterization of the topographic distribution pattern of peaks in cortical evoked responses yields a complementary value to EPs assessment and interpretation for clinic and research.

Keywords: somatosensory evoked potentials, Auditory Brainstem Responses, Intraoperative Neurophysiologic Monitoring, Intensive Care Unit, neuroplasticity

INTRODUCTION

Event Related Potentials (ERP) can be elicited by a wide variety of sensory, cognitive or motor events. They are thought to reflect the summed activity of postsynaptic potentials produced when a large number of similarly oriented cortical pyramidal neurons fire in synchrony while processing information.

ERP became particularly accessible as a noninvasive test in humans when those methods were developed in the 1940's. ERP allowed researchers to separate out the small evoked signals elicited by sensory stimuli from the larger voltage oscillations present in the spontaneous electroencephalogram. This advent of signal averaging, and subsequently the use of digital computers (Galambos and Sheatz 1962), promoted a rapid expansion in the application of ERP techniques. Currently, it is one of the most widely used methods in neuroscience research and clinical studies. Signal averaging has made it possible to record low-amplitude electrical potentials, in response to external stimuli, manifested as a wave or series of peaks that occurs in the order of thousands of milliseconds.

Here we will be referring to Evoked Potentials (EPs) that are recorded from the nervous system's electrical response to the stimulation of specific sensory pathways. ERP measures the electrophysiological response of the brain cortex and have the possibility of obtaining information relative to the functional integrity of pathways within the Nervous System (NS).

Evoked responses can be performed by multiple kinds of procedures and protocols to be applied depend on the symptoms, anamnesis, and other available information of the patient. These techniques have excellent temporal resolution and are important when the full cooperation of the patients is not possible. Depending on each case, neurophysiologists can use multiple bipolar or referential montage recording techniques. EPs recording for clinical use must preferably be bilateral. In general, all sensory modalities may be tested, although in routine clinical practice three types of evoked potentials are included: Auditory Brainstem Response (ABR), Visual Evoked Potentials (VEP), and Somatosensory Evoked Potentials (SSEP).

Since Jewett and Williston (1971) first described the ABR, it has become a valuable tool for probing auditory system function and development. The ABR is used in clinical practice to assess hearing thresholds. It is useful to identify deafness in neonates and infants whom conventional audiometry may not be useful. The major use of VEP is in the detection of subclinical lesions within the visual system, particularly in the diagnosis of demyelinating disease, and the SSEP have been used to study disorders of the brain, brainstem, spinal cord, sensory spinal nerve roots and peripheral nerves. Their clinical value is based upon their localizing role that is reinforced if there existconcordant neurological abnormalities. However, they may also detect subclinical lesions of the sensory pathways. In general, it is worth emphasizing that their results are not pathognomonic for any specific disease.

EPs recording are typically done with few channels, and it is recorded by surface electrodes on the skin or scalp. The interpretation includes assessing wave morphology, as well as measurement of latency and amplitude of the peaks, to detect deviations from the normal range (Charroó-Ruíz 2001; Chiappa and Hill 1997; Binnie 1995). For example, the SSEP are the complex series of peaks generated in somatosensory pathways at peripheral, spinal, subcortical and cortical levels (see Figure 1). They can be elicited by electrical or other stimuli, including tactile, mechanical and thermal, but in clinical practice the SSEP are elicited most effectively by electrical stimulation of a peripheral nerve, which preferentially activates the faster conducting fibers from proprioceptors and mechanoreceptors traveling in the dorsal lemniscal pathway.

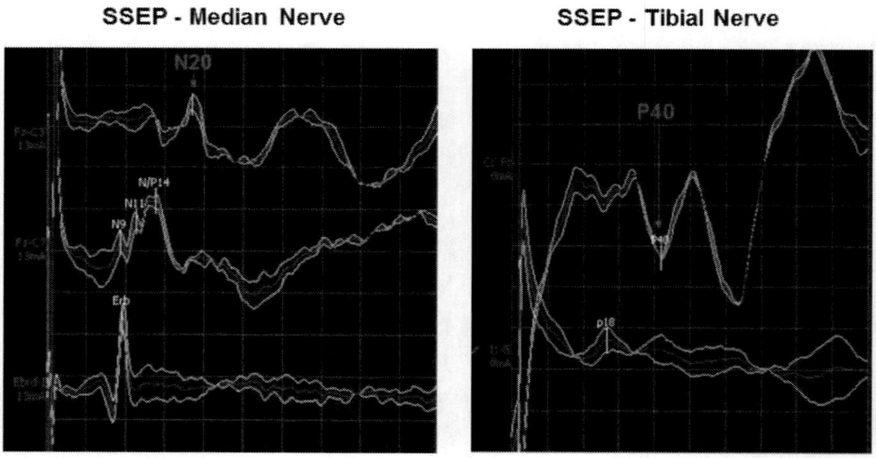

Figure 1. Typical SSEP waveform in healthy subjects.

EPs are reliable, easy to obtain, and objective. The within-subject reliability, stability and sensitivity of many evoked responses allow the technique to be used to evaluate changes in NS function (occurring in the course of diseases and in response to treatment). Changes in activity resulting in peak latency variation and/or amplitude modulation are considered abnormal, for example, clinical abnormalities may be reflected by latency prolongations.

EPs complement the neurological examination providing evidences that may sometimes be decisive in reaching a firm diagnosis. EPs have the advantage that can be recorded in patients who are anaesthetized or comatose. Furthermore, EPs can be used in cross-sectional and longitudinal studies. These studies are regularly performed at hospitals in the Clinical Neurophysiological Laboratory, although also can be performed for Intraoperative Neurophysiologic Monitoring (IONM) during surgery and at Intensive Care Unit (ICU).

EPs have multidisciplinary uses and are modified by different disorders, and intrinsic and extrinsic factors. The variability by extrinsic factors include, but is not limited to: the stimulus intensity and inter-stimulus interval, impedance of recording electrodes, the filtering and amplification of recorded signals. Specific parameters for eliciting and measuring SSEP have been established as recommendations for uniform testing conditions (Standards for short latency somatosensory evoked potentials 2006).

The purpose of this chapter is to describe neurophysiological changes that occur, immediate and at short-term evolution, in the structures of the NS in patients evaluated in three different contexts of work: IONM, ICU, and research in the Clinical Neurophysiological Laboratory.

INTRAOPERATIVE NEUROPHYSIOLOGIC MONITORING (IONM)

IONM of patients with lesion of the NS that need surgery is important to preserve the function on the NS. IONM, widely used in the lumbosacral spine surgery, can prevent damage of neural structures at risk and thus prevent the onset of a new neurological deficit during trans and postsurgical periods, which is supported by a large body of evidence published in the last decade (Deletis and Sala 2008; Sutter 2007a, 2007b; Paradiso 2006; Krassioukov 2004, Fehlings 2010; Sala 2007). The marginal extra cost of the IONM is minimal compared to the proved benefit.

Here are shown examples of uses at the IONM from our team for preservation of cranial nerve, nerve roots and spinal pathways, by assessing of patients (adults and children) with posterior fossa tumors, and injury of the cervical and lumbosacral column. IONM was made with a protocol adjusted according to type and extent of injuries. Different Electrodiagnosis (EDX) tests were used: Motor Evoked Potentials (MEPs), Electromyography (EMG), F Wave, and mapping of the nerve roots and cranial nerves. Synchronizing Neurosign-100 and Digitimer-D185 stimulators at the equipment Neuronica-5

(*Neuronic S.A.*) it was possible to evaluate the neural pathway and direct electrical stimulation of motor nerve, nerve roots.

During the IONM it was possible to detect the 50% amplitude decreases of the P40 peak (alarm signal) in SSEP obtained by electrical stimulation of the tibial nerve (see Figure 2).

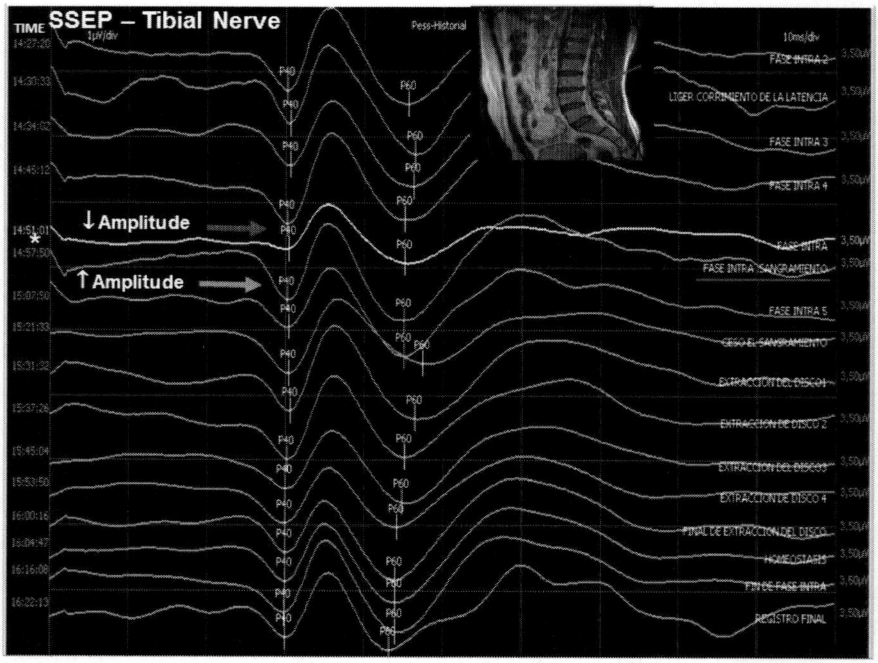

Figure 2. SSEP by tibial nerve stimulation and Magnetic Resonance Imaging. Baselines records show waveform clearly defined and consistent (first signal). Decrease amplitude of the responses was confirmed and the surgeon was notified changing his procedure. In few minutes the cortical SSEP return to normal amplitude. All records obtained after that were stable.

Paraparesis, paraplegia and quadriplegia are complications of spinal surgery and certain surgeries of the aorta. Nuwer (1995) describes that during spine surgery the incidence of major neurologic complications (paraplegia) is generally 1%. However, this condition is so devastating for the patient and family that justify doing the IONM systematically during surgery to prevent, the incidence of paraplegia, paraparesis, and other neurological complications.

We have seen also immediate improvement in electrophysiological responses that were interpreted as signals of good prognosis. Figure 3 shows the records of the tibial nerve SSEP performed in the operating room of a male patient, 56 years old, with a cauda equina syndrome and an extruded disc herniation at L4-L5, which is marked by an arrow on the Magnetic Resonance Imaging (MRI) study of the lumbosacral region. The first response during the IONM (baseline corresponding averaged obtain in the operating room at 12:51 pm), before beginning the surgery, showed the presence of P40 cortical response, of a very low amplitude and prolonged latency. However, note that at 13:31 pm, the response is becoming wider and therefore easier to identify the P40 peak, which continues during the remaining records made until the end of the surgery. The improvement in the morphology and amplitude of the response corresponded with the moment in which the neurosurgeon had already performed the decompression (laminectomy). Even the P40 peak latency, still prolonged, slightly decreases its value, 2 ms shorter latency.

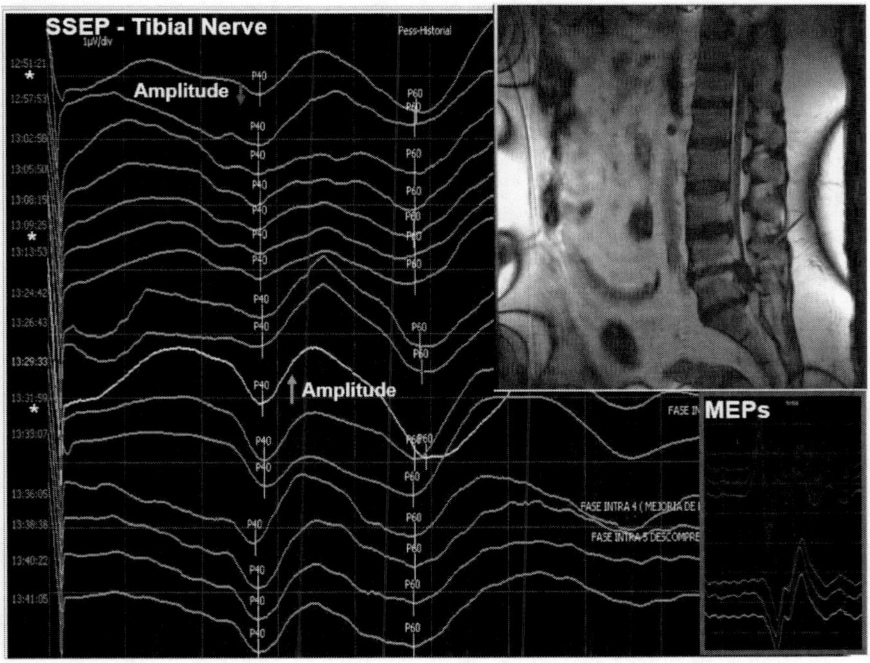

Figure 3. SSEP by tibial nerve stimulation (P40), MRI (up at the right of the figure) and MEPs (bottom at the right of the figure). Surgeon tells to the technologist that decompression is done. Cortical SSEP showed a waveform clearly defined and consistent. Note the increase of amplitude of P40 peak at the end of the surgery.

In some patients, it was possible to offer a favorable prognosis, in the trans-operative evolution based on the electrophysiological signal behavior during IONM, which were classified as immediate signs of good prognosis. In our IONM work experience, caught our attention, that in some patients the post-surgery outcomes could be predicted. In particular, when the response of P40 peak was initially absent during a successful surgery procedure it is possible to obtain it. In this sense, Kelleher et al. (2008) suggest that the recovery of SSEP following his initial loss is a good indicator of the absence of a new post-surgical neurological deficit. However, in general, few studies are published addressing the prognostic value of SSEP during IONM for functional recovery. They emphasize the alarm signals.

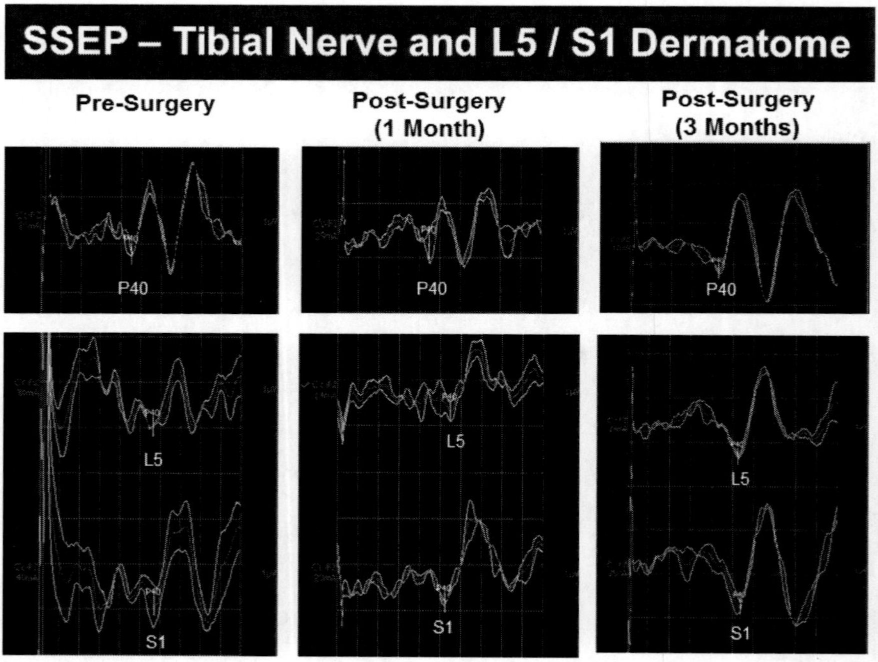

Figure 4. SSEP by stimulation of the tibial nerve and L5/S1 dermatomes. Note the changes in the cortical responses (improvement) in the post-operative evaluation (at one and three months) of operated patient by an L4-L5 disc herniation.

The information offered immediately by neurophysiologists during the IONM is of great value to the surgeon. Although there are modern imaging studies with high anatomical resolution, electrophysiological techniques are still very useful, because they provide vital information about the functional state of

NS with high time resolution (Domans 2010; Fehlings 2010; Kothbauer and Deletis 2010; Sutter 2007a, 2007b).

During the realization of the IONM both signals –alarm signals and good prognosis signals- were immediately notified to the neurosurgeon, while, postoperative short-term of the patients confirmed the preservation, and the improvement of function after surgery (see Figure 4).

Another positive aspect during IONM in clinical practice is to implement multimodal protocols. In example, the recording of MEPs by trans-cranial electrical stimulation, Free-run EMG, SSEP and ABR (see Figure 5). This step is important since it places the trans-surgical evaluation according to the standards established for IONM for spine surgery (Nuwer 2012; Domans 2010; Malhotra and Shaffrey 2010; Fehlings 2010; Kothbauer and Deletis 2010; Quraishi 2009; Kelleher 2008; Sutter 2007c; Paradiso 2006; Krassioukov 2004). It is required if you want to make an optimal monitoring of the functional integrity of the spinal cord, especially considering the complexity of some of the surgical techniques currently used. These results allow surgeons the possibility to get immediate feedback on the condition of the neural structure at risk during spinal surgeries.

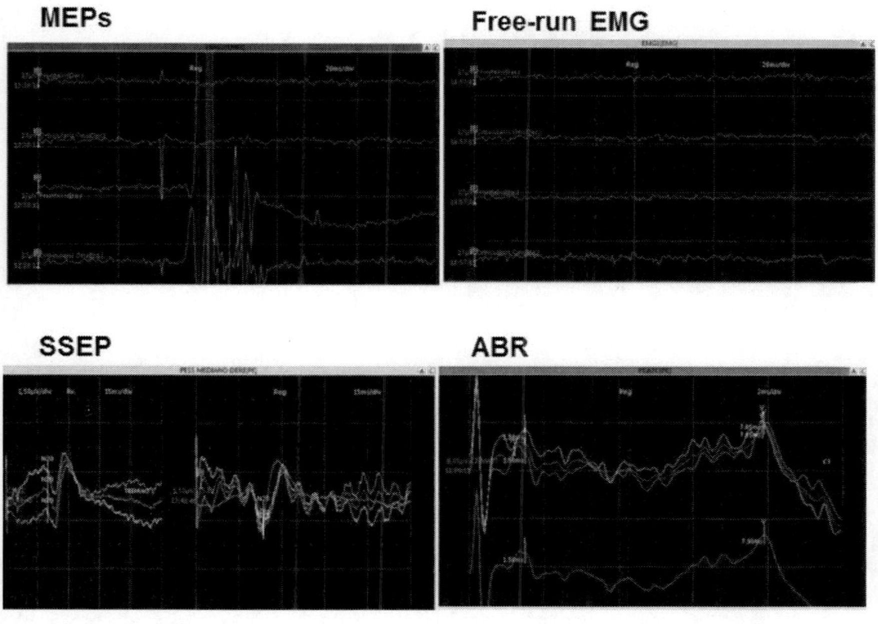

Figure 5. Multimodal protocol for IONM.

MEPs are electrical signals recorded from neural tissue or muscle after activation of central motor pathways. They complement other neurophysiological tests, such as SSEP, in the assessment of the NS, especially during IONM. Malhotra and Shaffrey (2010) noted that although MEPs monitoring is considered safe, relative contraindications include epilepsy, cortical lesion, skull defect, proconvulsant medication, cardiac pacing, and implantable devices. MEPs by transcranial electrical stimulation, ABR, Free-run EMG of cranial nerves, EEG, and SSEP may be considered medically necessary during spinal surgery, intracranial surgery, or vascular procedures.

In summary, neurophysiological tests provides functional quantified data on multimodal afferent and efferent pathways through the brain and long spinal tracks. It is possible to contribute to the preservation of the function of the nerve roots and spinal pathways, cervical and lumbosacral columns using an IONM protocol with EPs and different EDX tests during surgeries in adults and children. However, the recording protocols should follow certain standards. In all patients, one should be aware that its interpretation might be biased by several factors.

Intensive Care Unit

Neurocritical care is the intensive care provided to patients with severe neurological and neurosurgical conditions. To date, the evaluation of brain damage has been estimated using combinations of several indirect factors, such as neurological examination, Glasgow Coma Score (GCS) motor score, etc. However, the association between GCS motor score and neurologic outcome remains unknown; although the data show that initial GCS motor score examination can at least provide baseline objective prognostic data for decisions by healthcare professionals.

On the other hand, early prognosis is challenging, because sedation, ventilation, hypothermia and neuromuscular blockade potentially confound neurological assessment. The neurologist and neurophysiologist are often asked to assess patients at the ICU. We clearly need an accurate predictor to avoid withdrawal of treatment from those who might make a good neurological recovery.

Prediction of the prognosis in surviving comatose patients is still one of the unsolved problems in neurocritical care. Currently, the neurophysiological tests being used for predicting an outcome in patients at ICU include

Electroencephalogram (EEG) reactivity, SSEP, ABR, and different EDX tests, as EMG and NCS (Nuwer 2012; Guréit 2010; Jonghe 2008).

There are many types of EEG recordings that can be applied in the ICU. Conventional EEG, which is a gold standard for evaluating the neurologic function, only provides data over a period of 30 min and requires highly trained neurophysiologist to analyze it, while, continuous EEG (cEEG) monitoring (over 24 h) should be initiated within 1 h of seizures onset.

On the other hand, SSEP has been perhaps the most evaluated and heavily relied-upon tool for assessing the prognosis of patients in coma, especially after cardiopulmonary resuscitation (van Putten 2012; Guérit 2010). SSEP are fairly valuable for early prediction of the outcome of post-anoxic brain damage. Bilateral absence of response from the sensory cortex (N20) is the best predictor of failure to regain consciousness, i.e., death, a persistent vegetative state or minimal consciousness, with a high specificity, with an estimated false negative rate below 2% following anoxic ischemic injury (Zandbergen 1998). Similar levels of negative predictive power have also been found for patients with traumatic injuries (Robinson 2003; Sleigh 1999). Carter and Butt (2005), after a systematic review, report that although imperfect, SSEP appears to be the best single overall predictor of outcome. There is sufficient evidence for clinicians to use SSEP in the prediction of outcome after brain injury.

Neurophysiological test are used extensively and increasingly in the ICU to assess the degree of injury and predict the outcome for comatose patients (Kuroda 2016; Cruse 2014; Amantini 2008; Botteri 2008; Chiappa 1998). Although the prognostic value of the different clinical neurophysiological parameters varies as a function of the origin of coma, four general principles should be kept in mind, irrespective of cause. First, clinical neurophysiology only evaluates the current functional brain status, so that any prognostic information needs to be updated in case of eventual complications. Second, clinical neurophysiology may reflect brain dysfunction, irrespective of its pathophysiology, so that the prognostic value of any given degree of EEG or EPs alteration may vary as a function of the degree of reversibility or irreversibility of the underlying process. Third, whenever there is evidence of focal dysfunction, the electrogenesis corresponding to the most intact neural structure should be taken into account for general prognosis. Fourth, the preservation of ERP (for example, P300) predicts awakening with a very high probability, irrespective of coma etiology (Guérlt 2010).

We provide here evidence of the value of the EPs and EDX tests, in the assessment of patients in ICU with severe neurological damage _ due to infection by dengue. This has been observed more frequently in recent years.

We present the results of the neurophysiological evaluation of one female patient with a diagnosis of post-dengue encephalo-myelo-radiculo-neuropathy through studies, during the admission at the ICU, as well as the monitoring one year after. In this prospective study, we found many neurological complications with expressions in the neurophysiological studies. In all sections were performed EPs, EMG, and NCS with the equipment Neuronica-5 (*Neuronic S.A., Cuba*). From the findings in the neurophysiological records was possible to characterize the functional state of the NS structures in this patient, which was consistent with neurological deficits presented in the course of the disease. Early severe changes in electrophysiological recordings were illustrated at Figure 6. During the clinical recovery the electrophysiological responses were changing slowly over time (months), being possible to provide evidence of the disease course through the measured parameters in each electrophysiological signal.

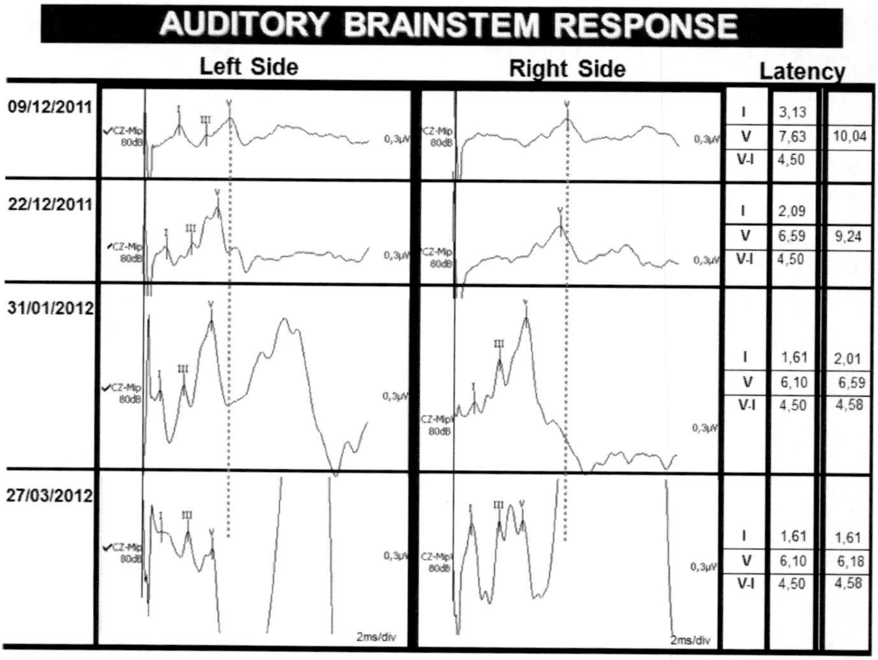

Figure 6. (Continued)

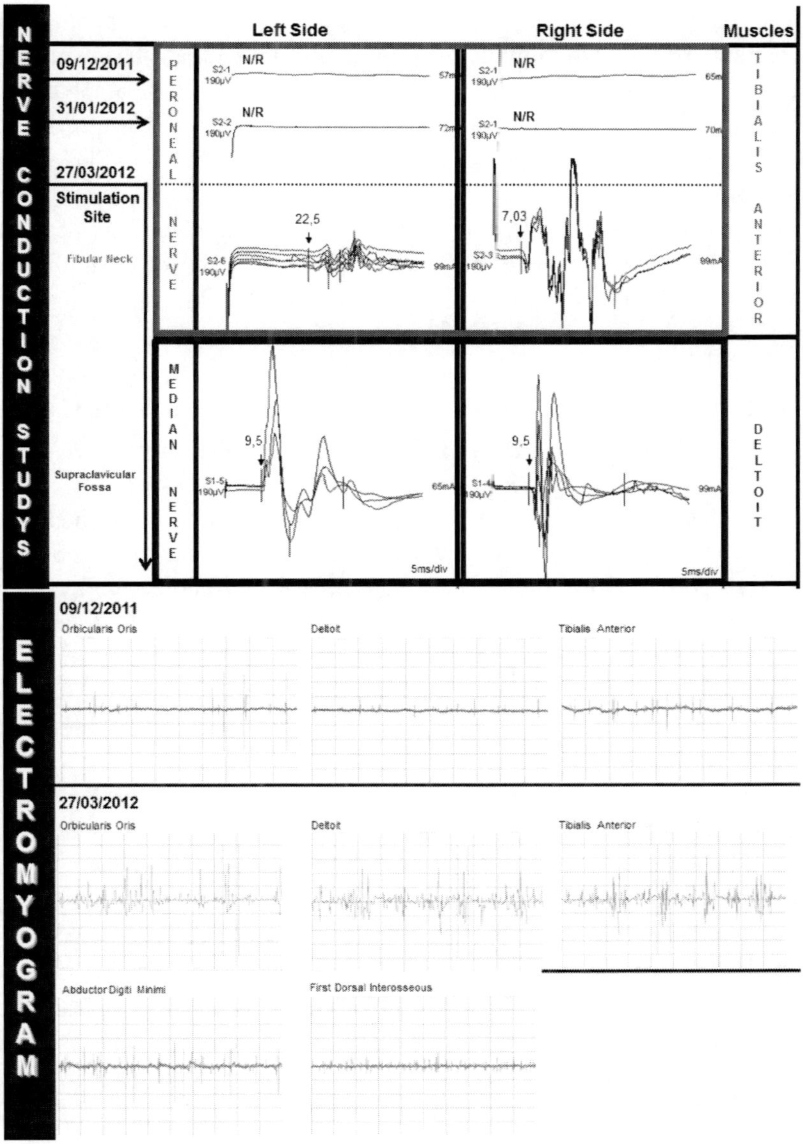

Figure 6. Neurophysiological tests in a patient with post-dengue encephalo-myelo-radiculo-neuropathy. (Box A) Auditory Brainstem Response. Note the increase of latency of I, III, and V waves on left side, and only for the V wave on the right side in the first study. (Box B) Nerve Conduction Study noted a loss of the responses in the first study. (Box C) Electromyogram records, only left derivation are shown. In general after several months all neurophysiological responses showed an improvement.

The optimal use of neurophysiological tests in the ICU is very important, and particularly, it substantially improves the individual management of patients. EPs have become a clearly established diagnostic tool supplementing the EEG in ICU. They should be considered only when there is a reasonable possibility that new and relevant information will be provided to the management of the patient.

In summary, it is possible to provide evidences of the NS functional state, by immediate changes in neurophysiological tests at ICU which is an important support for monitoring and prognosis of the patient in coma.

EPs as a Neurophysiological Technique to Investigate Neuroplasticity

Neuroplasticity is defined as the nervous system's ability to change and repair itself after an injury (Bavelier and Neville 2002). Plasticity in its broadest form, refers to the brain' ability to change, for neurons and networks to alter their function because of intrinsically or extrinsically factors. When some cortical regions do not receive an adequate sensory input, these brain regions become vulnerable to recruitment from other sensory modalities, and this form of plasticity is known as Cross-Modal Plasticity (Hauthal 2013; Shiell 2014; Lomber 2010; Finney 2001).

Investigation of human brain function is possible from recording activity of large-scale brain networks during EEG, Magnetoencephalography (MEG), functional Magnetic Resonance Imaging (fMRI), Positron Emission Tomography (PET) or Single Photon Emission Computed Tomography (SPECT). However, fMRI has been the most used tool in published research on neuroplasticity in subjects with single sensory deprivation, visual or auditory (Merabet and Pascual-Leone 2010; Bavellier and Neville 2002; Sadato 2002; Finney 2001). Numerous studies of neuroplasticity have documented findings on Cross-Modal Plasticity in the absence of sensory input such as hearing or vision, from both experimental animal models and human studies (Meredith 2011; Dye 2009; Sadato 2004, 2002, 1996; Finney 2003, 2001; Bavelier 2000; Cohen 1997; Neville 1987, 1983). To date, however, little is known about what happens when both sensory modalities are lost, deaf-blindness (Obretenova 2010; Osaki 2006, 2004).

The advent of electronic averaging, and then subsequently the use of digital computers (Galambos and Sheatz, 1962), promoted a rapid expansion in the application of EPs techniques. An advantage of EPs over neuroimaging techniques is that it is associated with relatively low cost for setup and maintenance. The stability and reproducibility of EPs is sufficient to allow their use to assess changes of the functional state of NS on time (Passmore 2014). Therefore, EPs can be useful in assessing neuroplastic changes that occur in normal course of development or to monitor changes in recovery of the brain function.

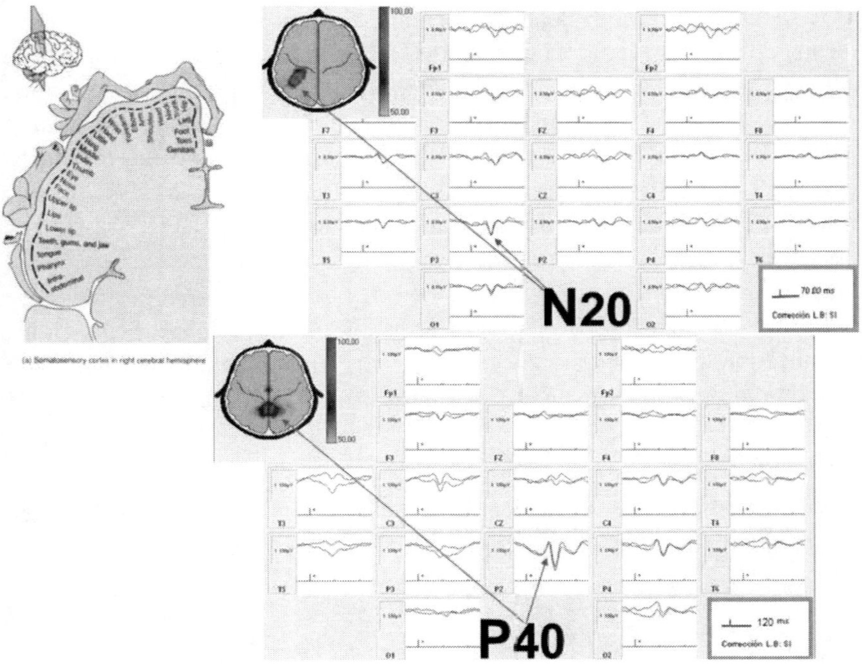

Figure 7. Cortical evoked responses of the SSEP (N20 and P0) from healthy children.

Figure 7 show SSEP in healthy normal-hearing children. Observe the replicability of the recording, an expression of good signal–noise ratio, and the typical morphology of the N20 and P40 peaks corresponding to the somatosensory cortex activation after stimulation of the median and tibial nerves, respectively. Following placement of 19 recording electrodes using the International 10-20 System, SSEP were recorded and then processed. Topographic maps were obtained for SSEP N20 (median nerve) and SSEP P40 (tibial nerve), permitting assessment of cortical reorganization. To evoke somatosensory response, the median and tibial nerves were stimulated on both sides separately with a Neuropack 2 (*Nihon Kohden Corporation, Japan*) electrical stimulator, coupled to the MEDICID electroencephalograph (*Neuronic S.A., Cuba*).The results obtained in healthy children are very important as normative data (Charroó-Ruíz 2012).

Electrophysiological descriptions of topographic distribution of somatosensory cortical response on the scalp have precedents in the studies of Penfield, who described somatotopic representation in the post central gyrus through direct stimulation of the cortex. Therefore, it is a very important gold standard for the interpretation of SSEP recording obtained from the scalp.

SSEP are useful for evaluating the central somatosensory pathway, even in patients with normal brain MRI findings. SSEP abnormalities can reflect the dysfunction of brain areas functionally connected to other regions of the cortex.

In deaf and deafblind children Cochlear Implant (CI) candidates, a significant change was found in SSEP N20 topography (hand representation area). There is over-representation in SSEP N20 topography, which is selective for the response obtained by right median nerve stimulation (see Figure 8), with an expansion of activated areas into the left temporal, parietal and occipital regions, normally intended for processing others sensory modalities (auditory input in the temporal region, visual information in the occipital region). This reorganization was greater in subjects with longer duration of hearing and vision loss and in those whose vision loss was very severe (total deaf-blindness) from an early age.

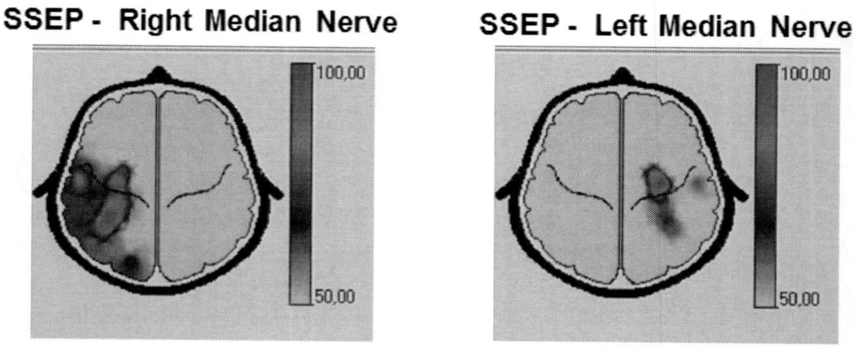

Figure 8. Topographic maps for cortical SSEP by stimulation of median nerve (N20 peak). In children with dual sensory deprivation the representation was asymmetrical, with more over-representation on the left hemisphere— to the evoked responses by stimulation of the right and left median nerves.

Researches using EEG and MEG methods in adults have demonstrated that somatotopic representations of the body in the human cortex can be mapped by examining the spatial pattern of evoked responses to somatosensory stimulation of different body parts. However, little attention has been paid to the development of somatotopic neural organization in infants (Saby 2015). The topography of SSEP cortical responses also has been relatively not much studied (Kakigi and Shibasaki 1992, 1991; Lesser 1987, Emerson and Pedley 1990, Charroó-Ruíz 2005).

The research presented here, evaluating cortical reorganization in deaf and deaf-blind children in comparison with healthy normal-hearing children (Charroó-Ruíz 2012) are the first evidences of Cross-Modal Plasticity using SSEP. Papers on deaf-blind children, in general, are cases descriptions, providing limited information (Filipo 2004; El-Kashlan 2001; Saeed 1998; Chute 1995).

Although vision and hearing are extraordinarily important senses in normal humans, our findings suggest that somatosensory information plays an important role in everyday communication in deaf and deaf-blind subjects. These children base their communication on tactile language and, therefore, cortical somatosensory representation expands, particularly in the contra-lateral hemisphere (area of her/his dominant hand), more in deaf-blind children (Figure 9).

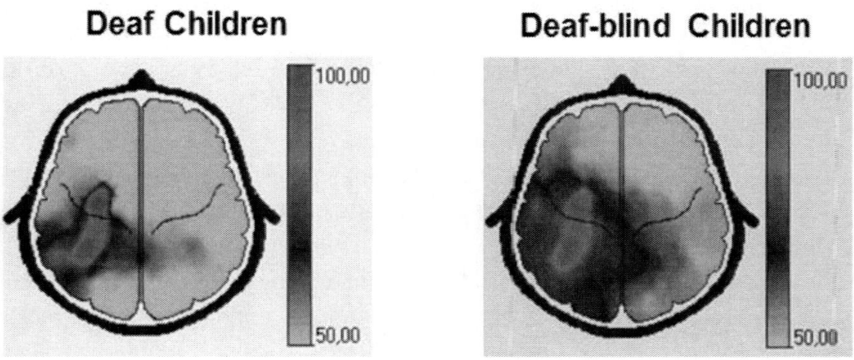

Figure 9. N20 SSEP topographic maps obtained by median nerve stimulation in deaf and deaf-blind children. N20 SSEP topography showed extended activation to other sites that normally process sensory input from other modalities, auditory in deaf children, and auditory and visual in deaf-blind children.

This study used electrical stimuli appropriate for the somatosensory pathway stimulation, and the one most used in clinical practice for neurophysiological assessment of conduction in the dorsal column pathway. We interpreted the change observed in N20 SSEP topography as a cortical reorganization of somatosensory representation, an expression of Cross-Modal Plasticity.

As the criteria for cochlear implantation in children expands, neurophysiological evidences of Cross-Modal Plasticity may play a critical role in determining whether implantation may be beneficial in non-typical pediatric patients, in example deaf-blind children. Therefore, nontraditional measures such as the topographic distribution of the N20 SSEP could be useful in assessing the neuroplastic changes in deaf and deaf-blind children, as these changes may directly influence outcomes following audiological intervention, such as CI.

In summary, this results expand the body of evidence on neuroplasticity in deaf and deaf-blind subjects, particularly in children CI candidates. However, research about neuroplasticity in deaf and deaf-blind children requires further study using neurophysiological tests, to assess if SSEP will be useful to describe changes during auditory rehabilitation post-CI.

Finally, although the data presented here represent evidence with case-studies these findings add to the growing body of evidence about neuroplasticity. EPs are established as a meaningful neurophysiological measurement technique in both clinical and research contexts, where prospective longitudinal studies provide important information concerning the course of the disease of the patient and in response to treatment. Specific parameters for eliciting and measuring EPs have been established as recommendations for uniform testing conditions.

Although the EPs and other neurophysiological tests are relatively simple and non-invasive, they require specially trained technicians and clinical neurophysiologists.

In the future, more efficient data processing techniques may achieve even more sensitive and specific evoked potential tests and hence, even more appropriate for diagnostics, for intraoperative, neurointensive monitoring, and neuroscience research.

Conclusion

EPs together with other neurophysiological tests provide a mean to noninvasively brain functioning assessment in patients at ICU and IONM. EPs are largely used for supplement information from clinics, imaging methods, and other laboratory tests. Nowadays SSEP and ABR are the most important EPs used, and show interesting findings in the patient assessment at IONM and ICU. Characterization of the topographic distribution pattern of peaks in cortical evoked response gives more value to EPs assessment and interpretation for use in clinic and research.

References

Amantini, A., Amadori, A. and Fossi, S. (2008) Evoked potentials in the ICU. *European J of Anaesthesiology,* 25:196-202.

American Clinical Neurophysiology Society. (2006) Standards for short latency somatosensory evoked potentials.

Bavelier, D. and Neville, H. (2002) Cross-modal plasticity: where and how? *Nat Rev Neurosci,* 3:443–52.

Bavelier, D., Tomann, A., Hutton, C., Mitchell, T., Corina, D., Liu, G. and Neville, H. (2000) Visual attention to the periphery is enhanced in congenitally deaf individuals. *J Neurosci,* 20:RC93.

Binnie, C.D., Fowler, C. and Cooper, R. (1995) Clinical neurophisiology: EMG, nerve conduction and evoked potential. Oxford-Boston: Butterworth-Heinemann. pp. 325-327, 359-360, 393-420.

Botteri, M. and Guarneri, B. (2008) Electrophysiological test in intensive care. *European J of Anaesthesiology,* 25:174-180.

Carter, B.G. and Butt, W. (2005) Are somatosensory evoked potentials the best predictor of outcome after severe brain injury? A systematic review. *Intensive Care Med,* 31:765-775.

Charroó-Ruíz, L., Aznielle, T., Borrego, M., Antelo, J.A., Díaz-Comas, L., Suárez, A., et al. (2005) Adquisición de la respuesta cortical del Potencial Evocado Somatosensorial del nervio tibial con múltiples derivaciones de registro. *Rev Ing Electrónica, Automática y Comunicaciones* 2005; 16:39–44.

Charroó-Ruíz, L., Hernández, R., Pérez-Abalo, M., Machado, C., Galán, L. and Aznielle, T. (2001) Potenciales Evocados. *Técnicas neurofisológicas y aplicaciones clínicas. Habana: Editorial Academia*. pp. 1–25, 77–114.

Charroó-Ruíz, L., Pérez-Abalo, M.C., Hernández, M.C., Álvarez, B., Bermejo, B., Bermejo, S., Galán, L. and Díz-Comas, L. (2012) Cross-modal plasticity in Cuban visually impaired child cochlear implant candidates: topography of somatosensory evoked potentials. *MEDICC Review*, 14:23–9.

Chiappa, K.H. and Hill, R.A. (1997) Short-latency somatosensory evoked potentials: Interpretation. In: Chiappa KH, editor. *Evoked Potentials in Clinical Medicine.* 3rd ed. New York: Lippincott-Raven. pp. 341–400.

Chiappa, K.H. and Hill, R.A. (1998) Evaluation and prognostication in coma. *Electroencephalography and Clin Neurophysiology*, 106:149-155.

Chute, P.M. and Nevis, M.E. (1995) Cochlear implants in people who are deaf-blind. *J Visual Impairm Blindness*, 89:297–301.

Cohen, L.G., Celnik, P., Pascual-Leone, A., Corwell, B., Falz, L., Dambrosia, J., Honda, M., Sadato, N., Gerloff, C., Catalá, M.D. and Hallett, M. (1997) Functional relevance of cross-modal plasticity in blind humans. *Nature,* 389:180–3.

Cruse, D., Norton, l., Gofton, T., Young, G.B. and Owen, A.M. (2014) Positive Prognostication from Median-Nerve Somatosensory Evoked Cortical Potentials. *Neurocrit Care,* 21:238–244.

Deletis, V. and Sala, F. (2008) Intraoperative neurophysiological monitoring of the spinal cord during spinal cord and spine surgery: a review focus on the corticospinal tracts. *Clin Neurophysiol,* 119:248-264.

Dormans, J.P. (2010) Establishing a standard of care for neuromonitoring during spinal deformity surgery. *Spine,* 35:2180-2185.

Dye, M.W., Hauser, P.C. and Bavelier, D. (2009) Is visual selective attention in deaf individuals enhanced or deficient? The case of de useful field of view. *PLoS One,* 4(5):e5640.

El-Kashlan, H.K., Boerst, A. and Telian, S.A. (2001) Multichannel cochlear implantation in visually impaired patients. *Otol Neurotol,* 22:53–6.

Emerson, R.G. and Pedley, T. (1990) Somatosensory evoked potentials. In: Daly DD, Pedley TA, eds. *Current practice of clinical electroencephalography.* 2nd ed. New York. Raven Press, 679-706.

Fehlings, M.G., Brodke, D.S., Norvell, D.C. and Dettori, J.R. (2010) The evidence for intraoperative neurophysiological monitoring in spine surgery. *Spine*, 35:S37-S46.

Filipo, R., Bosco, E., Mancini, P. and Ballantyne, D. (2004) Cochlear implants in special cases: deafness in the presence of disabilities and/or associated problems. *Acta Otolaryngol Suppl*, 552:74–80.

Finney, E.M., Clementz, B.A., Hickok, G. and Dobkins, K.R. (2003) Visual stimuli activate auditory cortex in deaf subjects: evidence from MEG. *Neuroreport,* 6:1425–7.

Finney, E.M., Fine, I. and Dobkins, K.R. (2001) Visual stimuli activate auditory cortex in the deaf. *Nat Neurosci,* 4:1171–3.

Galambos, R. and Sheatz, G.C. (1962) An electroencephalograph study of classical conditioning. *Am J Physiol,* 203:173-184.

Guérit, J.M. (2010) Neurophysiological testing in neurocritical care. *Current Opinion in Critical Care,* 16:98–104.

Hauthal, N., Sandmann, P., Debener, S. and Thorne, J.D. (2013) Visual movement perception in deaf and hearing individuals. *Adv Cogn Psychol,* 9:53–61.

Jewett, D.L. and Williston, J.S. (1971) Auditory-Evoked far fields averaged from the scalp of humans. *Brain,* 94:681-696.

Jonghe, B., Lacherade, J., Durand, M. and Sharshar, T. (2008) Critical Illness Neuromuscular Syndromes. *Neurol Clin,* 26:507–52.

Kakigi, R. and Shibasaki, H. (1991) Effects of age, gender, and stimulus side on the scalp topography of somatosensory evoked potentials following median nerve stimulation. *J Clin Neurophysiol,* 8(3):320–30.

Kakigi, R. and Shibasaki, H. (1992) Effects of age, gender, and stimulus side on the scalp topography of somatosensory evoked potentials following posterior tibial nerve stimulation. *J Clin Neurophysiol,* 9(3):431–40.

Kelleher, M.O., Tan, G., Sarjeant, R. and Fehlings, M.G. (2008) Predictive value of intraoperative neurophysiological monitoring during cervical spine surgery: a prospective analysis of 1055 consecutive patients. *J Neurosurg Spine,* 8:215–221.

Kothbauer, K.F. and Deletis, V. (2010) Intraoperative neurophysiology of the conus medullaris and cauda equine. *Childs Nerv Syst,* 26:247-253.

Krassioukov, A.V., Sarjeant, R., Arkia, H. and Fehlings, M.G. (2004) Multimodality intraoperative mpnitoring during complex lumbosacral procedures: indications, techniques, and long-term follow-up review of 61 consecutivo cases. *J Neurosurg* (Spine1), 3:243-253.

Kuroda, Y. (2016) Neurocritical care update. *J of Intensive Care,* 4:36.

Lesser, R.P., Luders, H. and Dinner, D. (1987) The source of "paradoxical lateralization" of cortical evoked potentials to posterior tibial nerve stimulation. *Neurology,* 37: 82-88.

Lomber, S.G., Meredith, M.A. and Kral, A. (2010) Cross-modal plasticity in specific auditory cortices underlies visual compensation in the deaf. *Nat Neurosci,* 13: 1421–1427.

Malhotra, N.R. and Shaffrey, ChI. (2010) Intraoperative electrophysiological monitoring in spine surgery. *Spine,* 35:2167-2179.

Merabet, L.B. and Pascual-Leone, A. (2010) Neural reorganization following sensory loss: the opportunity of change. *Nat Rev Neurosci,* 11(1):44– 52.

Meredith, M.A., Kryklywy, J., McMillan, A.J., Malhotra, Sh., Lum-Tai, R. and Lomber, S.G. (2011) Crossmodal reorganization in the early deaf switches sensory, but not behavioral roles of auditory cortex. *Proc Natl Acad Sci,* 108:8856–61.

Neville, H.J. and Lawson, D. (1987) Attentation to central and peripheral visual space in a movement detection task: an event-related potential and behavioural study. II. Congenitally deaf adults. *Brain Res*, 405(2):268–83.

Neville, H.J., Schmidt, A. and Kutas, M. (1983) Altered visual evoked potentials in congenitally deaf adults. *Brain Res*, 266(1):127–32.

Nuwer, M.R., Dawson, E.G., Carlson, L.G., Kanim L.E. and Sherman J.E. (1995) Somatosensory EP spinal cord monitoring reduces neurologic deficits after scoliosis surgery: results of a large multicenter survey. *Electroencephalogr Clin Neurophysiol,* 96:6-11.

Nuwer, M.R., Emerson, R.G., Galloway, G., Legatt, A.D., López, J., Minahan, R., Yamada, T., Goodin, D.S., Armon, C., Chaudhry, V., Gary S. Gronseth, and Harden, C.L. (2012) Evidence-Based Guideline Update: Intraoperative Spinal Monitoring with Somatosensory and Transcranial Electrical Motor Evoked Potentials. *J Clin Neurophysiol,* 29:101–108.

Obretenova, S., Halko, M.A., Plow, E.B., Pascual-Leone, A. and Merabet, L.B. (2010) Neuroplasticity associated with tactile language communication in a deaf-blind subject. *Front Hum Neurosci*, 4:3:60.

Osaki, Y., Doi, K., Takasawa, M., Noda, K., Nishimura, H., Ihara A, Iwaki, T., Imaizumi, M., Yoshikawa, T., Oku, N., Hatazawa, J. and Kubo, T. (2004) Cortical processing of tactile language in a postlingually deaf-blind subject. *Neuroreport*, 15(2):287–91.

Osaki, Y., Takasawa, M., Doi, K., Nishimura, H., Iwaki, T., Imaizumi, M., Oku, N., Hatazawa, J. and Kubo, T. (2006) Auditory and tactile processing in a postmeningitic deaf–blind patient with a cochlear implant. *Neurology,* 67:887–890.

Paradiso, G., Lee, G.Y.F., Sarjeant, R.B.S., Massicotte, E.M. and Fehlings, M.G. (2006) Multimodality Intraoperative Neurophysiologic Monitoring Findings During Surgery for Adult Tethered Cord Syndrome: Analysis of a Series of 44 Patients With Long-Term Follow-up. *Spine,* 31:2095–2102.

Passmore, S.R., Murphy, B. and Lee, T.D. (2014) The origin, and application of somatosensory evoked potentials as a neurophysiological technique to investigate neuroplasticity. *J Can Chiropr Assoc*, 58:170-183.

Quraishi, N.A., Lewis, S.J., Kelleher, M.O., Sarjeant, R., Rampersaud, Y.R. and Fehlings, M.G. (2009) Intraoperative multimodality monitoring in adult spinal deformity: analysis of a prospective series of one hundred two cases with independent evaluation. *Spine,* 34:1504–1512.

Robinson, L.R., Micklesen, P.J., Tirschwell, D.L. and Lew, H.L. (2003) Predictive value of somatosensory evoked potentials for awakening from coma. *Crit Care Med,* 31:960–7.

Saby, J.N., Meltzoff, A.N. and Marshall, P.J. (2015) Neural body maps in human infants: Somatotopic responses to tactile stimulation in 7-month-olds. *NeuroImage,* 118:74–78.

Sadato, N., Okada, T., Honda, M. and Yonekura, Y. (2002) Critical period for cross-modal plasticity in blind humans: a functional MRI study. *Neuroimage,* 16:389–400.

Sadato, N., Pascual-Leone, A., Gramfan, J., Ibañez, V., Deiber, M.P., Dold, G. and Hallett M. (1996) Activation of the primary visual cortex by Braille reading in blind subjects. *Nature,* 380(6574):526–8.

Sadato, N., Yamada, H., Okada, T., Yoshida, M., Hasegawa, T., Matsuki, K.I., Yonekura, Y. and Itoh, H. (2004) Age-dependent plasticity in the superior temporal sulcus in deaf humans: a functional MRI study. *BMC Neurosci,* 5:56.

Saeed, S.R., Ramsden, R.T. and Axon, P.R. (1998) Cochlear implantation in the deaf-blind. *Am J Otol,* 19(6):774–7.

Sala, F. (2010) Intraoperative neurophysiology is here to stay. *Childs Nerv Syst,* 26:413-417.

Sala, F., Faccioli, F., Lanteri, P. and Gerosa, M. (2007) Surgery for intramedullary spinal cord tumor: the role of intraoperative (neurophysiological) monitoring. *Eur Spine J,* 2(16 Suppl):S130-S139.

Shiell, M.M., Champoux, F. and Zatorre, R.J. (2014) Enhancement of visual motion detection thresholds in early deaf people. *PLoS ONE,* 9:e90498.

Sleigh, J.W., Havill, J.H., Frith, R., Kersel, D., Marsh, N. and Ulyatt, D. (1999) Somatosensory evoked potentials in severe traumatic brain injury: a blinded study. *J Neurosurg,* 91:577–80.

Sutter, M. (2007b) The diagnostic value of multimodal intraoperative monitoring (MIOM) during spine surgery: a prospective study of 1,017 patients. *Eur Spine J,* 16:162–170.

Sutter, M. (2007c) The validity of multimodal intraopeative monitoring (MIOM) in surgery of 109 spine and spinal cord tumors. *Eur Spine J,* 16:197 – 208.

Sutter, M., Eggspuehler, A., Muller, A. and Dvorak, J. (2007a) Multimodal intraoperative monitoring: an overview and proposal of methodology based on 1,017 cases. *J Spine Eur,* 16(Suppl 2):S153–S161.

Van Putten, M. (2012) The N20 in post-anoxic coma: Are you listening? Clinical Neurophysiology. *Clin Neurophysiol,* 123(7):1460-4.

Zandbergen, E.G., de Haan, R.J., Stoutenbeek, C.P., Koelman, J.H. and Hijdra, A. (1998) Systematic review of early prediction of poor outcome in anoxicischaemic coma. *Lancet,* 352:1808–12.

In: Evoked Potentials and Electrical Stimulation ISBN: 978-1-53611-075-3
Editor: Timothy Eisenhower

Chapter 2

CORTICAL-EVOKED POTENTIALS: MIGRAINE, STROKE AND RELATED PSYCHIATRIC COMPLICATIONS

Massimiliano Toscano*, MD, PhD,
Alessandro Viganò, MD, PhD
and Vittorio Di Piero, MD, PhD
"Sapienza" University of Rome,
Department of Neurology and Psychiatry, Rome, Italy

ABSTRACT

In the last years, cortical-evoked potentials have been extensively used to study any abnormal cortical information processing associated with a specific neurotransmission pathway (i.e., serotonergic neutrasmission), which could represent a key element of several neurological conditions.

In this view, the intensity dependence of auditory-evoked potentials (IDAP), and the habituation deficit of visual and auditory-evoked potentials (VEP and AEP habituation deficit, respectively), represent the two main cortical evoked paradigms adopted in neuropsychiatric research fields.

* Corresponding Author: Massimiliano Toscano, MD, PhD, Department of Neurology and Psychiatry, "Sapienza" University of Rome, Viale dell'Università 30, 00185 Rome, E-mail: massimiliano.toscano@uniroma1.it.

IDAP is thought to be inversely proportional to the central serotonergic tone; thus, it represents a useful non-invasive indicator of serotonergic activity associated with specific clinical and behavioural correlates. For this reason, many studies investigated central serotonergic tone alteration in several neuropsychiatric disease.

On the other hand, the habituation phenomenon of neural evoked activity represents a physiological decrement in the behavioural response that results from repeated stimulation and that does not involve sensory adaptation or fatigue. A deficit of habituation (i.e., reduction or absence of the normal decrease in cortical response after repeated sensorial stimulation) is thought to be a core feature of several neurological conditions, in particular of migraine, where lack of habituation (or even an amplitude increase in the cortical response) has been observed in the interictal phase with various stimulus modalities.

In the present chapter, we will focus on the importance of cortical-evoked potentials to analyse the neurophysiological correlates of migraine, stroke and related psychiatric complications as well, paying particular attention to the possible implications in terms of clinical applications and future researches.

INTRODUCTION

In the last years, cortical-evoked potentials have been extensively used to study any abnormal cortical information processing associated with a specific neurotransmission pathway (i.e., serotonergic neutrasmission), which could represent a key element of several neurological conditions. Namely, the neuropsychiatric research field entrusted with clinical use of cortical-evoked potentials has mainly aimed to search for a neurophysiological correlate of neural monoamine pathways involved in migraine, stroke and in psychiatric post-stroke comorbidities as well.

One of the first adopted paradigms was the P300 wave analysis of auditory-evoked potentials, which is an event related potential (ERP) found during the performance of various cognitive tasks [1]. On one hand, P300 amplitude reduction was found in severe endogenous depression [2], which is a pathophysiology mainly related to serotonin tone alteration; on the other hand, changes in P300 latency were investigated in stroke patients as well[3]. Interestingly, an increase in P300 latencies with normal amplitudes were found in case of a single hemispheric cerebral infarction [4,5], regardless of the post-stroke impairment degree. This finding was further confirmed by studies that found an increase in the P300 ERP component latency in stroke patients with minor functional impairment, namely those with lacunar stroke [6] and those with minor stroke [7].

Given the neurophysiological and neuroanatomical characteristics of the P300 ERP component, it has been suggested that a single ischaemic lesion may alter dopaminergic and noradrenergic cortical-subcortical pathways, therefore generating or modulating the ERP located in the tempo-parietal cortical areas [7, 8].

Notwithstanding, P300 analyses have been poorly used to analyse neurophysiological correlates of neuropsychiatric diseases; this is mainly due to the fact that P300, as well as the other long latency event related components, is strongly influenced by cognitive aspects, therefore making any correlations with a specific monoamine neural pathway almost doubtable.

In this view, the intensity dependence of auditory-evoked potentials (IDAP), and the habituation deficit of visual and auditory-evoked potentials (VEP and AEP habituation deficit, respectively), represent the two main cortical evoked paradigms adopted in neuropsychiatric research fields.

INTENSITY DEPENDENCE OF AUDITORY-EVOKED POTENTIALS (IDAP)

IDAP rationale directly came from the augmenting-reducing (A-R) paradigm, firstly described by Buchsbaum and Silverman [9]. A-R is a measure of sensory inhibition mechanisms based on the reliable and individual differences in the way sensory evoked potentials change (i.e., change in amplitude). The original paradigm otherwise suffered from several technical problems, mostly concerning anatomical structures generating the observed evoked potentials as well as AEP reproducibility. These problems were overridden by two fundamental aspects modified in the original paradigm; first of all AEPs, which reflect activity from the superior temporal plane and include the auditory cortex, are definitely preferred compared to VEPs, which otherwise are thought to reflect activity of a non-specific centro-parietal cortex. Second, Scherg and von Cramon [10] applied the dipole source analysis to the original A-R paradigm. Namely, two temporal sources were identified bilaterally for the N1/P2 component of AEP: a tangential dipole representing activity in the superior temporal plane in which the primary auditory cortex is situated, and a radial dipole reflecting the evoked activity of the lateral part of the superior temporal gyrus corresponding to secondary areas [11]. On one hand, the test–

retest-stability of N1/P2 components for AEPs was shown to be higher when measured by the activity of the dipole source [12]. On the other hand, the application of dipole source analysis allowed researchers to differentiate the primary auditory cortex with its high serotonergic innervation from secondary auditory areas.

This was fundamental for the IDAP paradigm to be developed, since it allowed for the correlation of this paradigm with the central serotonergic tone. This is mostly because the primary sensory cortices layer IV is most densely innervated by serotonin and in turn receives specific thalamic sensory input, mostly from dorsal raphe nuclei (DRN) characterised by a very regular and stable discharge (i.e., pacemaker properties). Therefore, as initially reported by Hegerl et al. [13], a high firing rate of the serotonergic neurons in the DRN, which generates a high serotonergic neurotransmission, is associated with a small increase in the evoked cortical response with increasing loudness of the stimuli (or, in other words, with a weak IDAP). A low central serotonergic neurotransmission, on the other hand, is related to a strong IDAP, therefore suggesting the intriguing hypothesis that IDAP is inversely proportional to the central serotonergic tone.

In the first years since IDAP was described, its role as an indicator for central serotonergic activity was criticised by the study from Dierks and colleagues, since they found that the tryptophan depletion test did not sufficiently increase IDAP. Notwithstanding, this test has been challenged as a reliable tool to study central serotonergic activity [11, 14]. Moreover, several biochemical and neurophysiological studies confirmed a role of serotonergic neurotransmission in the modulation of AEP intensity dependence. In fact, a negative correlation between the concentration of 5-HIAA in the CSF and IDAP was subsequently described [15]. Strobel and co-workers [16] provided further evidence for a role of serotonergic neurotransmission in the modulation of the IDAP by demonstrating that a functional polymorphism in the promoter region of the serotonin transporter gene (5-HTTLPR) is associated with AEP intensity dependence. Moreover, a recent population-based study [17] also suggests that IDAP, along with BDNF polymorphisms, can predict altered serotonergic activity.

Finally, many studies described IDAP as a valuable tool for studying both migraine (reflecting the serotonergic tone alteration described in this pathology) and CNS effects of drugs acting on 5-HT transmission (see below in the chapter).

Therefore, IDAP is now widely considered as a reliable indicator of central serotonergic function, so that a high IDAP value (i.e., an increased linear amplitude/stimulus intensity function (ASF) slope) means a lower central serotonergic tone, and vice versa.

HABITUATION TO REPEATED SENSORY STIMULATIONS

The definition of habituation is a decrement in the behavioural response that results from repeated stimulation and that does not involve sensory adaptation or fatigue [18].

Habituation is a multifactorial event, inducing synaptic plasticity while being considered as a fundamental adaptive behaviour of the nervous system and being involved in learning and memory [19].

Habituation is a basic form of brain learning. From the phylogenetic point of view, it is ubiquitous and appears to be highly conserved in living organisms, from the simplest to the more complex. E.R. Kandel [20] described habituation to electrical stimulation in his famous *Aplysia* model in a series of works. For this reason, it is occasionally considered as the more simple form of learning. Although habituation also represents a pivotal form of neuronal plasticity, its neural mechanism remains mostly unknown.

A reason why habituation is so strongly conserved in living organisms is that it somehow performs a pre-learning function, filtering out low-salient sensory stimuli in order to allow brain resources to focus on important or new stimuli.

In the past, habituation has been defined by several names (i.e., acclimatization, accommodation, negative adaptation, fatigue, etc.), so it is quite difficult to identify when this mechanism has been initially studied. However, the process of sensory response habituation is independent from sensory adaptation and motor fatigue. The main feature that allows scientists to discriminate between the habitation-mediated and fatigue-mediated decrement of responses is the phenomenon of dishabituation. Dishabituation is defined as a rapid recovery of the previous habituated responses due to the occurrence of an unexpected stimulation that is different from the ongoing stimulation. Apart from dishabituation, other methods allow us to distinguish between habituation and fatigue in an experimental setting where dishabituation cannot be studied [21].

In his paper on the history of habituation, Thompson [22] reported that the seminal experiment aimed at evoking a habituation response was done by Hernández-Peónand and colleagues in 1956. Authors described the developing of habituation to an auditory click stimulation in the auditory system of an unanaesthetized cat. In brief, monotone clicks were delivered at 0.5 Hz for an adequate time span and authors recorded a progressive reduction of evoked responses in the cochlear nucleus (the first relay of the auditory system). Although the ability to habituate the cochlear nucleus was lately controverted, this was the first time in which the habituation of sensory responses was investigated.

Habituation influences responses at both neural and behavioural level, including hormone release. Apart from behavioural responses that can be directly observed, neuronal activity may be measured by resting of the evoked EEG or functional neuroimaging as a progressive decrement of event-related potentials or BOLD signal measured by functional magnetic resonance imaging [23]. To date, a clear description of molecular and cellular mechanisms of habituation is not known.

Habituation is a process that is present in all sensory modalities of various types of stimuli, mainly in the Central Nervous System (CNS) at all neuronal relays, from the first-order sensory nucleus to the sensory cerebral cortices.

Habituation is characterised by some fundamental parameters described by Thompson and Spencer in 1966[24] that are lately revised by a pool of experts in 2009[18]. The original and revised characteristics of habituation are given in Table 1. (A detailed comment of their aspects lies outside the purpose of this chapter, but a precise discussion may be found in Thompson [22] and in Rankin et al. [18]).

In brief, (1) a habituation-dependent decrement of responses can be modelled as both linear and exponential; (2) habituation can be reverted spontaneously at the end of the stimulus; and (3) the habituation degree is steeper for higher stimulation frequency and weaker intensity.

Habituation presents stimulus specificity (a non-habituated response is elicited for stimuli other than the habituating one) as well as a frequency-dependent spontaneous recovery modality (i.e., recovery is faster after high frequency stimulation than lower frequency stimulation).

Among explicative models of habituation, three major theories have been proposed. In this chapter, we will focus on the Groves and Thompson's Dual Process Theory [25], since most of the experiments made on habituation in neurological diseases have been framed in this theory. However, it is worth at least mentioning the other two theories: 1) The Stimulus-Model Comparator Theory by Eugene Sokolov (1960) and 2) the Gnostic Hypothesis presented by Konorski and revised later by Allan Wagner. The deep analysis of these theories lies outside the purpose of this chapter, but interested readers may find a comprehensive description in Thompson et al. [22].

According to the dual-process theory by Groves and Thompson, any effective sensory stimulus will start two independent and opposite processes in the central nervous system. One of them is a progressive decrement of brain responses (i.e., habituation) and the other is an increment of responses (i.e., sensitization). Habituation and sensitization probably rely on different brain mechanisms and develop in separate brain regions, although they interact to yield the final behavioural outcome of the stimulus-response pathway. Sensitization is an elementary form of behavioural plasticity as well. In this model, sensitization is mostly active at the beginning of the stimulation when the stimulus is new and highly salient. With the maintaining of the stimulation pattern, the saliency content reduces and habituation develops. In this model, habituation is a result of stimuli repetition. According to Groves and Thompson's dual-process theory, the habituation of responses depends on a synaptic transmission decrement that can be qualified as a memory process [26].

It is worth mentioning that the dishabituation process differs from the pathological one that is known as a "lack of habituation" (which we will be described in detail later), since dishabituation relies on sensitization to sensory stimuli, which is due to a heterosynaptic facilitation mechanism [20].

It was confirmed by Eric Kandel and his colleagues in the 1970s that dual-process theory was linked to the genus *Aplysia*. *Aplysia* is a a genus of gastropod mollusk in which monosynaptic responses are easily evoked by noxious stimulation. In this model, Kandel found that habituated responses to repeated stimulation depended on a decreased probability of neurotransmitter release at the level of the sensory-motor synapse. This was mostly due to a decrease of calcium ion influx at the presynaptic level.

A limitation of the dual-process theory in comparison to both Sokolov's and Wagner's theories is that the dual-process theory focuses on short-term habituation (within session habituation), while the other two take into account long-term habituation. To date, an open debate surrounds the concept of long-term habituation. In the revision paper, Rankin et al. [18] stated that long-term habituation cannot be considered as an extended form of short-term habituation, but rather as a form of long lasting memory for habituation training, based on different cellular and molecular mechanisms than those involved in short-term habituation, building a frame that appears particularly knotty [27-29].

Table 1. Original and revised characteristics of habituation as described by Groves and Thompson (1970), and Rankin (2009), respectively

Characteristics	Original (Groves and Thompson, 1970)	Revised (Rankin et al. 2009)
1	Given that a particular stimulus elicits a response, repeated applications of the stimulus result in decreased response (habituation). The decrease is usually a negative exponential function of the number of stimulus presentations	Repeated application of a stimulus results in a progressive decrease in some parameter of a response to an asymptotic level. This change may include decreases in frequency and/or magnitude of the response. In many cases, the decrement is exponential, but it may also be linear; in addition, a response may show facilitation prior to decrementing because of (or presumably derived from) a simultaneous process of sensitization.
2	If the stimulus is withheld, the response tends to recover over time (spontaneous recovery)	If the stimulus is withheld after a response decrement, the response recovers at least partially over the observation time ("spontaneous recovery")
3	If a repeated series of habituation training and spontaneous recovery are given, habituation becomes successively more rapid (this phenomenon might be called "the potentiation of habituation")	After multiple series of stimulus repetitions and spontaneous recoveries, the response decrement becomes successively more rapid and/or more pronounced (this phenomenon can be called "the potentiation of habituation").
4	Other components being equal, the more rapid the frequency of stimulation, the more rapid and/or more pronounced the habituation.	Other components being equal, more frequent stimulation results in more rapid and/or more pronounced response decrement, and more rapid spontaneous recovery (if the decrement has reached asymptotic levels).
5	The weaker the stimulus, the more rapid and/or more pronounced the habituation. Strong stimuli may yield no significant habituation.	Within a stimulus modality, the less intense the stimulus, the more rapid and/or more pronounced the behavioural response decrement. Very intense stimuli may yield no significant observable response decrement.
6	The effects of habituation training may proceed beyond the zero or asymptotic response level.	The effects of repeated stimulation may continue to accumulate even after the response has reached an asymptotic level (which may or may not be zero, or may result in no response at all). This effect of

Characteristics	Original (Groves and Thompson, 1970)	Revised (Rankin et al. 2009)
		stimulation beyond asymptotic levels can alter subsequent behaviour, for example, by delaying the onset of spontaneous recovery.
7	Habituation of a response to a given stimulus exhibits stimulus generalization to other stimuli.	Within the same stimulus modality, the response decrement shows some stimulus specificity. To test for stimulus specificity/stimulus generalization, a second novel stimulus is presented, and a comparison is made between the changes in the responses to the habituated stimulus and the novel stimulus. In many paradigms (i.e., developmental studies of language acquisition) this test has been improperly termed as a dishabituation test rather than a stimulus generalization test, which is its proper name.
8	Presentation of another (usually strong) stimulus results in the recovery of the habituated response (dishabituation).	Presentation of a different stimulus results in an increase of the decremented response to the original stimulus. This phenomenon is termed "dishabituation." It is important to note that the proper test for dishabituation is an increase in response to the original stimulus and not an increase in response to the dishabituating stimulus (see point #7 above). Indeed, the dishabituating stimulus by itself need not even trigger the response on its own.
9	Upon repeated application of the dishabituatory stimulus, the amount of dishabituation produced habituates (this phenomenon might be called "the habituation of dishabituation").	Upon repeated application of the dishabituating stimulus, the amount of dishabituation produced decreases (this phenomenon can be called "the habituation of dishabituation").
10 (additional)		Some stimulus repetition protocols may result in properties of the response decrement (i.e., more rapid rehabituation than the baseline, smaller initial responses than the baseline, smaller mean responses than the baseline, and less frequent responses than the baseline) that last hours, days or weeks. This persistence of habituation aspects is termed "long-term habituation".

Migraine: Evidence from Cortical-Evoked Potentials

In the following paragraphs, we will primarily focus on migraine rather than other forms of primary headaches. This is mainly because most of the studies performed focused on migraine, and secondly because tension-type headache and cluster headaches seem to represent an extreme condition, in which habituation is only limitedly impaired (tension-type headache) or greatly altered (cluster headache) [30].

Habituation in Primary Headaches: Its Role for Migraine Pathophysiology

A lack of the physiological habituation to repeated sensory stimulation appears to be the neurophysiological hallmark of migraine [31,32], and it is considered the most reproducible abnormality in migraine interictally [33], albeit some studies didn't retrieve the same result, possibly due to methodological, genetic, environmental, psychological, and psychodynamic and geographically differences between studies (for a detailed review, see Magis et al.) [34].

During the mid -1980s, the lack of habituation had been firstly described in migraineurs by Schoenen and colleagues by using cognitive evoked potentials; latterly, this alteration has been found to influence all sensory modalities (visual, auditory, and somatosensory cortices).

In addition, the absence of normal habituation in migraineurs was found in blink reflex studies [35-37] as well as in laser evoked potentials (LEPs) studies, involving both the early and the late components generated in the secondary somatosensory cortex and in insular/anterior cingulate cortices, respectively [37-40].

Initially, this habituation deficit was associated to cortical hyperexcitability. Single and repeated transcranial magnetic stimulation used to solve the issue resulted in univocal data [41-45].

However, accumulating data showed that two different patterns seem to interplay. First, initial brain responses to a short sensory stimulation are smaller than in healthy controls. Second, reduced decrement or an increase in amplitude is observed for longer stimulation. The fact that VEPs' responses to a low number of stimuli are not increased, but rather decreased contrasts with the theory of cortex hyperexcitability. On the other hand, habituation seems to be lacking in migraineurs with respect to healthy subjects in apparent disagreement with the hypothesis of a hypoexcitable brain.

Interestingly, VEP habituation seems negatively correlated to the amplitude of the initial response to stimuli; that is, the higher the initial response, the higher the habituation, and vice versa [33, 46].

A possible explication is that brain cortex is hyper-responsive in migraine due to a reduced pre-activation level of the brain [47]. This hypothesis is based on the "ceiling theory", according to which habituation starts when sensitization causes evoked responses to reach an upper threshold [48].

VEP studies showed that fast components of VEPs within the EEG gamma band (30-80 Hz), called gamma band oscillations (GBOs), represent the interaction between thalamic and cortical areas [49]. Namely, early GBOs reflect thalamic transmission while late GBOs are typically due to cortex activity. In migraineurs, late GBOs show a lack of habituation, especially in the late components, suggesting that the main brain alterations of migraine are in the cortex. In 2011, Chen et al. [50] used MEG to study visual evoked fields (VEFs) and described a difference in VEFs between migraineurs and healthy volunteers, interictally. In agreement with the lack of habituation theory, episodic migraineurs showed a decreased first response and enhancement of field activity during this time due to stimulus repetition.

During auditory potentials, similar results have been described. Habituation is lacking in migraineurs when a sensory gating paradigm has been used [51]. Some results were provided using three blocks of 40 averaged responses delivered at low (10dB) and high (70 dB) intensities. In both cases, a lack of habituation has been found in migraineurs with respect to HV when the amplitude of an N1-P2 response has been monitored over time. Moreover, Ambrosini et al. [51] found that also the initial responses of N1-P2 are lower in migraineurs than in HV. This fact supports the data obtained by VEPs and strengthens the hypothesis that the main brain cortex alteration in migraine is a low preactivation level, and so it consequently experiences hyperexcitability. In addition, somatosensory evoked potentials have been used to investigate the lack of habituation. Relevant insights have been provided by the analysis of High Frequency Oscillations (HFOs). Like GBOs in visual-evoked potentials, HFOs represent the fast activity embedded within the classical broadband recording of a somatosensory potential, whose origin is basically in fronto-parietal cortices [52]. Intriguingly, HFOs seem not to undergo habituation while low frequency oscillations do [53]. This could imply that HFOs are involved in the transmission of the stimulus and prevent overload that is divided into two components, which are early HFOs and late HFOs. Early HFOs are thought to reflect spike activity in thalamo-cortical cholinergic fibers, while the late HFOs seem to represent cortical inhibitory GABA-ergic interneurons activity. In migraine, the habituation of the early HFO mechanism seems to be selectively impaired while late HFOs are normal. Additional evidence has been recently provided by the fact that the amplitude of early HFOs was significantly correlated to the clinical evolution; namely, spontaneous worsening was associated with reduced HFO amplitude while a spontaneous improvement was associated with enhanced HFO amplitude [54]. This dissociation suggested that the thalamo-cortical activity is decreased while cortical inhibitory activity (due to interneurons) is unaffected [55]. Taken together, these data suggest that lack of habituation is a phenomenon which influences predominantly the cortico-thalamic binding. Moreover, this supports the hypothesis that deficient habituation in migraine is due to a low pre-activation level, depending on the reduced thalamic control. Further evidence showed also that intracortical lateral inhibition of sensory cortices (the other possible culprit of the lack of habituation) is higher in migraine during the initial part of the stimulation. However, it progressively decreases with the progression of the stimulation. This phenomenon does not occur in heathy controls, where lateral inhibition remains somewhat stable during the stimulation. Lower inhibition and reduced preactivation level may thus also coexist, since the latter can promote the former via the reduction of lateral inhibition [30, 47].

Early HFOs' component deficits also strengthens the idea that interactions between the subcortical and cortical region could determine the pain free interval-attack migraine cycle. This interaction between subcortical and cortical regions is known as "thalamocortical dysrhythmia".

The experimental evidence of a habituation deficit at the cortical and subcortical levels can be combined in a unifying theory known as "thalamocortical dysrhythmia".

Further data on habituation theory may be derived from the migraine cycle and chronic migraine. Other experiments showed how in the pre-ictal phase (just before the attack), the amplitude and the habituation of some electrophysiological responses (i.e., CNV, P300) can become maximal [56, 57]. During the ictal phase, paradoxically, the habituation is restored and migraineurs react to stimuli as healthy subjects [58].

Chronic migraine is a complication of migraine, in which headache attacks affect more than 15 days during a month and at least 8 of them present migraineous features. Among risk factors, the commonest one is the medication overuse that is diagnosed in up to 80% of patients attending specialized headache clinics as Medication Overuse Headache (MOH) [59]. Interestingly, patients with chronic migraine (CM) have a neuromagnetic response pattern indistinguishable from that of EM patients during an attack, namely a high initial response amplitude and normal habituation. This pattern has been deemed a "never-ending attack" [60]. More importantly, when CM patients are successfully treated and the migraine frequency decreases, returning to an episodic migraine pattern, the VEPs also change to become similar to a pure episodic migraine pattern, which namely features low initial response and lack of habituation [61]. In a cohort of patients with MOH, Coppola and colleagues found that chronic patients with concomitant MOH had a particular neurophysiological pattern that differed from the episodic as well as pure chronic migraine: in CM, the early response is as high as control subjects while habituation is normal, resembling a typical ictal pattern. On the other hand, patients with MOH had an elevated initial response and a deficient habituation (as is seen with the pre-ictal pattern) [59].

LACK OF HABITUATION THEORY IN MIGRAINE: STATE OF THE ART

Recently, the pros and the cons of the lack of habituation theory has been reviewed in some extended reviews [30, 34].

The theory has been supported by several new studies that investigated several aspects of the lack of habituation. Cosentino et al. [62] found that cortical excitability and metaplasticity measured by MEPs change according to migraine cycle and the chronicization process. These changes may depend on the dysfunction of the glutamatergic regulation system in interictal patients with migraine without aura, which influences long-term depression and potentiation mechanisms [63]. Altered lateral inhibition was also described in the somatosensory cortex between attacks with a normalization during the attack [64].

The thalamocortical dysrhythmia seems to be a fluctuating process, changing over time according to the migraine cycle phase and burden of the disease, namely the chronicization process and disease duration. These changes are not only functional, but they can also involve both structures and connections between brain areas. A direct correlation between cutaneous pain threshold and the time for next migraine attack to occur was described by Schwedt and colleagues [65].

Using resting state fMRI, an abnormal network connectivity between the thalamus and visual cortex was associated with significant differences in the baseline thalamic microstructure [64].

From the structural point of view, interictally, patients with migraine without aura had a significantly lower gray matter density in the right inferior parietal lobule, right temporal inferior gyrus, right superior temporal gyrus, and left temporal pole compared to the controls. During an attack, the gray matter density increases in the left temporal pole, almost restoring normal values. Moreover, increased gray matter density was found in bilateral insula and right lenticular nuclei [66].

Stroke and Related Psychiatric Complications: Evidence from Cortical-Evoked Potentials

Stroke

Clinical use of cortical-evoked potentials in stroke patients is gaining popularity of the past few years. This is mostly because it is now widely accepted that a stroke lesion could lead to an alteration in the monoamine intracerebral level by affecting the cortico-cortical and cortico-subcortical monoamine circuits.

In this point of view, the first evidence of noradrenergic and dopaminergic neurotransmitter system involvement came from studies aimed at addressing the effects of ischaemic stroke on the P300 ERP component of auditory-evoked potentials.

Gummow et al. [3] described a decreased P300 amplitude in patients with middle cerebral artery stroke without any change in the latency; a more pronounced P300 amplitude decrease in those patients with a right hemispheric stroke was also reported. Onofrj and colleagues found a significant increase in the P300 latency in patients with thalamic haemorrhage [67], while P300 amplitude was normal. Similar P300 wave alteration (i.e., increase in the latencies with normal amplitudes) were hereafter confirmed also in the case of a single hemispheric cerebral infarction [4, 5], regardless of the post-stroke impairment degree. This hypothesis was further corroborated by studies that found an increase in the P300 ERP component latency in stroke patients with minor functional impairment, namely those with lacunar stroke [6] and those with minor stroke [7].

Given the neurophysiological and neuroanatomical characteristics of the P300 ERP component, it has been suggested that the increase in its latency, as shown in stroke patients, already suggests that a single ischaemic lesion may alter cortical-subcortical neurotransmitter pathways, namely the dopaminergic and noradrenergic ones, thereby generating or modulating the ERP located particularly in the tempo-parietal cortical areas [7, 8].

This research field took a pivotal turn when a growing amount of evidence started to highlight the importance of the serotoninergic system in stroke; this role was firstly suggested in pharmacological studies aimed at indirectly demonstrating a serotonergic modulation on brain excitability and motor recovery in stroke patients [68, 69]. Patients treated by the serotonergic drug showed a significant improvement in neurological status (as measured by the NIHSS), and a significant decrease of the motor cortex excitability with effects on both the affected and unaffected hemispheres, also associated with better motor recovery after stroke.

More recently, direct evidence came from studies which demonstrated central serotonergic pathways involvement after stroke by using cortical-evoked potentials [70, 71], namely the intensity dependence of auditory-evoked potentials (IDAP).

Although these studies were designed to investigate the serotonergic correlation with depression and aggressive behaviour in the acute phase of stroke, both of them reported an increased value of IDAP in all stroke patients as a group, which means that there is a decreased central serotonergic tone.

It is intriguing that the intensity dependence of AEP in stroke patients increased regardless of stroke severity, along with other stroke or patients' characteristic (i.e., past medical history, medication intake, ischaemic or haemorrhagic stroke, the site of the lesion, etc.). Hence, this finding further supports the hypothesis of direct involvement with the serotonergic system in the acute phase of stroke, with several implications in terms of stroke recovery and treatment as well.

Emotion, Mood and Behaviour after Stroke

Motor, sensory and cognitive deficit are widely considered as the most relevant deficits following stroke. Notwithstanding, changes in mood and behaviour after stroke represent a relatively unrecognised and undertreated group of post-stroke complications that affect as many as half of stroke patients, both in the very acute and chronic phases of the disease, also dramatically affecting post-stroke recovery and quality of life.

In this view, stroke represents a privileged disease for understanding the anatomical and biochemical basis of human behaviour; the first systematic study on emotional changes following cerebrovascular disease was conducted by Gainotti and colleagues in 1972 [72]. Starting from this innovative study, this research field achieved a growing interest over the last five decades.

Post-stroke depression (PSD) and aggressive behaviour after stroke are probably the most important mood and emotional disturbances, followed by post-stroke-fatigue (PSF), post-stroke anxiety and post-stroke emotional incontinence (PSEI).

From a pathophysiological point of view, it is still debated whether they represent an adaptive/reactive response to the physiological conditions that arise from stroke disability, or the direct effect of the stroke lesion, which may cause a disruption of cortico-cortical and subcortical-cortical monoamine circuits.

As a matter of fact, stroke entails a physiological response to a new disability and life-threatening event as well; to support this "reactive mechanism" hypothesis, a close relationship between PSD and post-stroke functional deficits has been highlighted, so that PSD could be a sort of adaptive reaction to these deficits [73, 74].

Notwithstanding, there is actually a great amount of evidences that support the biochemical basis underlying the emotional and mood disorders after stroke, regardless of the role of stroke disability and functional impairment. First of all, mood disorders seem to be more frequent after stroke than other physical illnesses with similar physical disability traits[75]; second, PSD has been described even after stroke with poor functional impairment (i.e., TIA and minor stroke) [76, 77], thereby suggesting a biochemical basis underlying stroke related psychiatric complications.

Most consisting findings that support a biochemical basis actually come from a biochemical study particularly focused on post stroke depression and on aggressive behaviour following stroke, the two main psychiatric comorbidities investigated with cortical-evoked potentials. The evidence resulting from this research is described in the following paragraphs, along with cortical-evoked potentials. As previously mentioned, these potentials are actually based on the hypothesis of a monoamine cerebral circuit of post-stroke disruption.

Post-Stroke Depression

Depression is the most common psychiatric post-stroke disease, affecting approximately one-third of stroke survivors both in the acute and chronic stages of the disease [75, 78]; a recent metanalysis shows that 31% of stroke patients developed depression at same points in time up to five years after stroke [79], with relevant implications in terms of stroke recovery and quality of life.

Even though the DSM-IV classification implies that strokes "cause" depression thorough a direct biological mechanism, the nature of the mechanism linking stroke and depression is still disputed. It has been debated for several decades to be a psychosocial mechanism, according to which the social and psychological stressors associated with a stroke are considered the primary cause of depression [80]. The "reactive theory" of post-stroke depression was somehow corroborated, since the evidence supporting a biological mechanism for post-stroke depression has not been consistently replicated. Moreover, two different metanalysis [81, 82] failed to identify any association between post-stroke depression and lesion location (i.e., left anterior lesions or left hemispheric lesions), and so several doubts concerning a direct stroke mechanism leading to depression have been realised.

Nowadays, even if social and psychological stressors associated with a stroke are thought to play an important role in the developing of PSD, depression has been reported to be more common after stroke than in other physical illnesses with similar levels of physical disability [83]. Moreover, patients with TIA, minor stroke, or even those with post-stroke anosognosia (i.e., those who are unaware of their disability, and therefore not able to experience a psychological reaction to their stroke) are frequently described to develop post-stroke depression [84-86]. Thus, the existence of a primary biological mechanism, according to which ischaemic insults directly affect neural circuits involved in mood regulation is widely accepted.

Indirect evidence from biochemical and pharmacological studies first indicated serotonin as the main monoamine involved in the biological mechanism underlying PSD. It is reported that a higher methylation status of the serotonin transporter gene promoter is associated with PSD between two weeks and one year after stroke, as well as a worsening of depressive symptoms over the first year after stroke [87]. Many authors already described selective serotonin recapture inhibitors (SSRI) as the more effective treatment for depression after stroke [88, 89], also advising their use even in the very early stage of the disease. Hence, the psychiatric research field highlighted the role of cortical auditory-evoked potentials as an indirect marker of depressive syndromes, namely predicting symptom development and response to SSRI treatment as well [90].

In this view, Vandoolaeghe and colleagues [91] investigated the P300 ERP component alteration in patients with major depression syndrome; they described a significantly higher P300 latency in depressed patients than in others. Moreover, among the whole group of patients with major depression, patients who did not respond to antidepressant therapy had a significantly higher pre-treatment of P300 latency than responders did. Gallinat et al. [92] reported that patients with a strong pre-treatment of LDAEP (an IDAP equivalent adopted in the context of psychiatric research) had a significantly greater decrease of depressive symptoms than those with a flat LDAEP; thus, a strong LDAEP, indicating low serotonergic activity, is related to a favourable response to acute SSRI treatment in depression. Similarly, another study demonstrated an important role of cortical-evoked potentials in predicting a clinical response to Citalopram [93]: a strong intensity dependence of the auditory-evoked potential was associated with a favourable response to SSRI treatment in a group of acutely depressed in-patients treated exclusively with Citalopram. The most interesting finding is that the role of auditory-evoked potentials in predicting clinical responses to SSRI seems not to reflect the temporary depressive state; in fact, neither Citalopram treatment nor a decrease of the total depressive symptoms scale score during treatment had a significant effect on IDAP in the patients' sample.

Thus, those psychiatric studies gave direct evidence of the central serotonergic tone's role in depressive syndromes. This made several researchers wonder if the clinical use of cortical-evoked potentials could also be useful in the neurological research fields, namely to investigate if a disruption of central serotonergic pathways involved in mood regulation, directly caused by the ischaemic stroke lesion, could represent the biological mechanism underlying post-stroke depression.

Yanay [94] reported changes in auditory P300 in patients with major depression and silent cerebral infarction. As expected, P300 amplitudes were significantly smaller in depressed patients than in normal controls. Then, in depressed patients, the P300 was re-examined after antidepressant treatment. It is intriguing that the changes in P300 amplitude after depression treatment already depended on the presence of silent infarcts. Patients without silent stroke showed an increase in P300 amplitude following a recovery from depression, whereas those with silent infarcts did not; in other words, the presence of silent stroke may interrupt the treatment-related P300 amplitude increase in depressed patients. This datum contributed to clarify that the link between stroke and depression is more than a simple reactive mechanism, therefore suggesting that a stroke lesion, even silent, may interrupt those neural monoaminergic pathways that are crucial for depression recovery, and perhaps for its development as well. This last hypothesis was later corroborated by a study from Rocco and colleagues [70], and aimed to mainly evaluate the serotonergic tone in the acute phase of stroke by using IDAP. It would then analyse the correlation between central serotonergic tone and depressive symptoms development in the very acute phase of stroke. To avoid depression diagnosis bias, they used the Montgomery and Asberg Depression Rating Scale (MADRS) [95] to analyse the presence of depressive symptoms (DS) instead of post-stroke depression, which is a diagnosis that is almost questionable in the acute phase of stroke. It is also remarkable that, to minimize the effect of possible confounders on IDAP results, they studied stroke patients with isolated subcortical infarcts. Interestingly, the intensity dependence of auditory-evoked potentials was more pronounced in those stroke patients with DS, which means a decreased central serotonergic tone, without any relationship between the occurrence of DS and the stroke severity and the site of the lesion. Taken together, this data suggests that the disruption of cortical-subcortical serotonergic pathways following stroke could play a more pivotal role in the development of DS than a single "strategic" lesion, regardless of the degree of disability after stroke.

This is in line with biochemical and pharmacological evidence that described serotonin as the monoamine primarily involved in the biological mechanism underlying depression development [87-89]. Despite the fact that both biochemical and neurophysiological studies highlighted the role of cortical and subcortical neural serotonergic pathways in the development of post-stroke depression, many studies also suggest an involvement of other monoamine circuits. Indirect evidences from pharmacological studies support the role of norepinephrine and acetylcholine in the pathogenesis of depression after stroke, as well as the role of alteration of dopaminergic pathways leading to changes in the amine level [96].

Hence, the pathophysiology of PSD is complex and likely involves the disruption of different monoamine neural pathways, as well as psychosocial and biological factors (i.e., a high postdexamethasone cortisol level, lower serum BDNF levels, overall brain perfusion reduction following stroke, etc.) [75, 97].

Anger and Aggressive Behaviour Following Stroke

Even if the pathophysiology of post-stroke depression actually involves different monoamine neural pathways in addition to the serotonergic one, the latter seems to play the most important role in the development of anger and aggressive behaviour after stroke; it is found to be present in 15-30% of patients, even in the acute stage of the disease [71, 98-100]. In the "Lausanne Emotion in Acute Stroke Study"[98], anger, together with aggression (which is considered to be its behavioural aspect), were observed in 23% of acute stroke patients during the first four days, and 13% of patients reported anger during the first seven days.

Kim et al., who studied stroke patients with anger and aggressive behaviour, described the central feature of these symptoms as an inability to control anger and aggression after stroke (ICAA); it is reported that at least 30% of stroke survivors experience ICAA in the first days following stroke [25].

The importance of this post-stroke psychiatric comorbidity lies in the bidirectional link between anger and stroke: on one hand, trait anger is associated with an increased risk of incident stroke [101], and on the other hand, anger and aggression are commonly observed in the first days of acute stroke, crucially affecting both clinical status and recovery. This could finally lead to worsened patient management and rehabilitation [102], as well as a stressful situation for the family, health care providers and patients [103].

Several studies designed to investigate the neurobiological basis of anger and aggression highlighted how the serotonergic system is involved in the modulation of offensive aggressive behaviour. Concentration levels of both serotonin (5-HT) and its main metabolite, 5-hydoxyindoleacetic acid (5-HIAA), were originally described to be lower in aggressiveness with respect to non-aggressive subjects[104]; moreover, reduced CSF concentration of 5-HIAA was demonstrated in individuals with a history of aggression[105] and impulsive aggression[106]. Also, pharmacological and neurochemical studies implicated the serotonergic processes in the regulation of different kinds of aggressive behaviour [107]: the prolactin elevation in response to a single dose of serotonin agonist was found to be lower in aggressiveness [108] and antisocial personality disorder [109].

From a neurophysiological point of view, many authors primarily focused on the inefficient sensory gating system in self-reported impulsive aggression as assessed by visual-evoked potentials [110]. A decreased amplitude of the P1 component seems to reflect a deficiency in the sensory gating system, thereby resulting in inefficient stimuli processing in aggressive subjects [111]; moreover a shorter P1, N1 and P2 peak latency was also described. Hence, it has been speculated that these data may indicate a quicker orienting and processing of stimuli to compensate low resting arousal level [110].

Interestingly, the most consistent findings in subjects who show aggressive behaviour is the so called evoked-potentials augmenting; augmenting-reducing (A-R) is a measure of sensory inhibition mechanism, based on the reliable individual differences in the way evoked potentials change (i.e., change in amplitude) with different stimulus intensities [110]. As mentioned before, the rational of A-R is the same of the intensity dependence of auditory-evoked potentials (IDAP), which is a measure of the changes in amplitude of evoked cortical response with increasing loudness of the stimuli. Hence, both IDAP and A-R are inversely related to the central serotonergic tone. In this view, it is intriguing that the augmenting of evoked potentials (i.e., a decreased serotonergic tone) is closely related to the disinhibition aspects of sensation seeking behaviour, and to impulsive aggressive behaviour as well [112]. This is evidence of the serotonergic pathway's involvement in the aggressive behaviour based on cortical-evoked potentials. More recently, a study supporting this evidence was also performed in regards to the acute phase of stroke [71]. The aim of this study was to evaluate the serotonin system's involvement in acute subcortical stroke patients with anger and aggression using the intensity dependence of auditory-evoked potentials. When grouping stroke patients together according to the presence of aggressive behaviour (AB), expressed by the overall Spielberger Trait Anger Scale (STAS) score, those with anger had a significantly steeper ASF slope (i.e., a higher IDAP value) with respect to stroke patients without anger and control. Since it is assumed that IDAP is inversely related to serotonergic tone, this datum seems to confirm an early decrease in serotonergic tone also in stroke patients with AB. Moreover, this is in line with previous studies that demonstrated reduced CSF concentration of 5-HIAA in individuals with a history of aggression [105] and impulsive aggression [106]. Interestingly, the early decrease in serotonergic tone was independent from stroke severity and the degree of functional impairment, since only patients with a mild motor deficit and NIHSS score lower than 7 were selected.

Thus, both biochemical and neurophysiological studies suggest that the serotonergic system is thought to have an inhibitory control over aggressive impulse; this contributed to define the so-called "serotonergic hypothesis of aggression," which considers the central serotonergic tone as inversely related to aggression. This dogma, even widely accepted for several decades, seems obsolete in the present day [113]. More specifically, whether the serotonin system has the same role both in the anger trait (TA) and in a state condition of anger (SA) is still debated.

In this view, Coccaro et al. [114] postulated that normal offensive aggression is positively related to serotonergic neuronal activity, whereas an inverse relationship probably exists between serotonin activity and impulse-like violent aggression. Van Der Vegt et al. [115] also reported a positive correlation between the level of trait-like aggression and basal CSF concentration of 5-HT and 5-HIAA, suggesting that trait and state aggression are probably differentially regulated by the central serotonergic system.

An important contribution to this debate was delivered by the aforementioned study, which aimed to evaluate the involvement of IDAP with the serotonin system in acute stroke patients with anger and aggression [71].

Despite the whole group of stroke patients with aggressive behaviour showing a higher IDAP value (i.e., a decreased central serotonergic tone) by grouping stroke patients according to the presence of the anger trait (TA) or a state condition of anger (SA), they found an opposite 5HT feature. On one hand, patients with TA, similarly to those with AB, showed a significantly increased IDAP value that is a decreased 5HT central tone. On the other hand, in sharp contrast to the hypothesis that the activity of a 5-HT system is inversely related to aggression, they found a significantly lower IDAP value in stroke patients with SA, suggesting an increase of central serotonergic tone. This means that an early decrease in serotonergic tone may be related to anger and aggression (representing a trait anger), without reflecting any change in aggressiveness after stroke (i.e., a state condition of anger). This confirms the doubt on the classic hypothesis of aggression being based on 5HT deficiency, which was already raised from previous reported biochemical studies.

Future Researches and Perspectives

The lack of habituation theory seems very promising not only because it explains the mechanisms of migraine pathophysiology, but also because pharmacological treatments of migraine are effective only in half of the cases, and they may have serious side effects; the most common used drugs are antiepileptics (valproate, topiramate), beta-blockers (atenolol), calcium-blockers (flunarizine), and tricycle antidepressants [116]. In addition, CM and MOH patients are refractory to therapy [117].

Beta-blockers and topiramate are able to normalize habituation in migraineurs [37, 118].

For this reason, non-invasive techniques of neurostimualtion, which are known as transcranial magnetic stimulation (TMS) and transcranial direct current stimulation (tDCs), have been applied to treat migraine patients. A recent meta-analysis of such studies has been published. The conclusion of this chapter conveys that, although there is a limited number of studies using such techniques in migraine, tDCs seems to be effective at least against pain intensity [119].

The main limitation concerning the majority of these studies is that they rarely investigated the effect of neurostimualtion paired with a neurophysiological assessment of brain responses. Hence, both excitatory and inhibitory stimulations have been used in migraine patients. Moreover, no distinction has been made between chronic and episodic migraineurs, albeit the lack of habituation theory suggested that these conditions might be extremely different from the neurophysiological point of view.

In a paper that combined neurophysiological recordings of habituation in episodic patients and tDCs treatment, an improvement of migraine was associated with a normalization of habituation at least immediately after the administration of the stimulation [120].

For the future, a more detailed study of brain responses to evoked potentials investigating the lack of habituation may be used to monitor the neurophysiological changes in a sample of patients undergoing pharmacological and neurostimulation intervention; this could otherwise clarify the action mechanism of these techniques that may be more complex than the simple increase or decrease in brain excitability.

Finally, a significantly reduced habituation of visual-evoked potentials has also been observed in a healthy subject with analytic information processing style, a specific cognitive behaviour characterised by a high activation of attention and continuous evaluation of stimuli [121]. This finding is somehow intriguing because the analytic information processing style has been associated with migraine [122]. Hence, further researches should be undertaken regarding the relationship between analytic cognitive style and lack of habituation in other physiological and pathophysiological conditions.

Concerning evoked potentials' clinical use in the neuropsychiatric research field, the main goal of future researches is to find any neurophysiological marker of mood disorders, which could allow the neurologist to choose the correct treatment depending on a single patient's biological and neurophysiological characteristics.

In this view, further researches are needed mainly because, beyond post-stroke depression and aggressive behaviour after stroke, several post-stroke psychiatric comorbidities deserve to be investigated by means of cortical-evoked potentials. This is the case with post-stroke emotional incontinence (PSEI), which is reported to be one of the most frequent post-stroke psychiatric comorbidities following stroke, with a prevalence ranging from 6 to 34% in the acute/subacute phase of stroke [78]. PSEI is characterised by excessive and inappropriate crying or laughing without apparent motivating stimuli, or in response to stimuli that would not normally evoke such responses; as reported by neuroanatomical and clinical studies [78, 123], the serotonin system is also thought to be the most involved in this psychiatric post-stroke comorbidity. In fact, PSEI was described to occur in patients with lenticulo-capsular and brainstem stroke, which therefore involve serotonergic fibers that ascend from the brainstem raphe nuclei to limbic forebrain structures, and then project through the basal ganglia to the frontal cortex. Serotonin involvement in PSEI development was also suggested by studies carried out in stroke patients, showing that serotonin transporter binding ratios in the midbrain and pons regions are lower in patients with PSEI [124]; serotonin gene polymorphisms are also described in these patients [125]. Taken together, these evidences gave the biological basis for SSRI treatment to be suggested as the most effective in reducing the frequency and severity of PSEI [126].

Otherwise, it is intriguing that dopamine and glutamine are also thought to be involved in PSEI pathogenesis. Moreover, the balance between glutamatergic excitation and inhibition is in turn modulated by other neurotransmitter systems, such as dopamine, serotonin and acetylcholine. Thus, it could be interesting to investigate, by means of cortical-evoked potentials, whether other antidepressant medications beyond those acting on the serotonergic system might be the appropriate treatment for a subgroup of patients with PSEI. There is also interesting evidence that makes this research field more intriguing: it is well-known that PSEI is strongly linked to PSD, a treatment that is described to potentially depend on the different monoamine pathways involved.

In this view that, even if in the last decades several researches were conducted on post-stroke depression, several questions remain unanswered.

First of all, it has been observed that only a subset of patients is responsive to SSRI treatment, while other patients respond to tricyclic antidepressants or MAO inhibitors; this treatment may not have a favourable profile concerning safety and tolerability in post-stroke patients.

Rampello et al. [127] suggested that this might depend on the existence of two distinct subsets of depression following stroke, as it is already described for primary depression. The first form, defined by the author as "retarded depression", actually refers to a state of clinical depression in which patients appear slow to initiate action and show anergia, hypokinesia and hypomimia; the other form, which is defined as "anxious depression", is mainly characterised by symptoms such as anxiety, insomnia and hostility.

Without a doubt, the existence of different subsets of post-stroke depression, which are also responsive to different treatment strategies, is the main reason why several studies reported conflicting results when trying to address the best treatment for post-stroke depression. Despite different recent metanalyses [128, 129] that suggest a beneficial effect of antidepressants on remission and response, there are still insufficient trials of each of the antidepressants to conduct meta-analyses of antidepressants thus far.

In this view, the real novel finding is that patients affected by anxious depression are responsive to SSRI, while those showing a ''retarded'' form of PSD show a good response to a noradrenaline reuptake inhibitor (i.e., Reboxetine) [127]. Hence, the most interesting aim of this research could be to address whether different clinical forms of PDS are responsive to different treatments because of a different underlying biochemical basis. In other words, in order to choose the correct treatment, it should be very helpful to correlate which form of PSD is responsive to NARI, and which one is responsive to SSRI, and pair them to a selective impairment of noradrenergic and serotonergic neural pathways, respectively.

The research from Linka and colleagues aimed to investigate whether IDAP could represent a valid differential indicator of serotonergic versus noradrenergic antidepressant therapy. In a primary study, the author reported that a strong intensity dependence of the auditory-evoked potential (i.e., a lower central serotonergic tone) was associated with a favourable response to SSRI treatment in a group of acutely depressed in-patients treated exclusively with Citalopram. Then, in another study from the same group, a highly significant correlation between lower intensity-dependent N1 amplitude slopes prior to Reboxetine treatment is reported, along with a stronger decrease in the Hamilton Depression Rating Scale.

Taken together, these studies suggest that those patients who showed a lower pre-treatment of a central serotonergic tone (i.e., higher IDAP value) were more likely to have a good clinical response to SSRI treatment, whereas those with a higher central serotonergic tone (i.e., lower IDAP value) were more likely to have a good clinical response to NARI treatment. It is also worth noting that neither applied antidepressants nor a decrease of the HDRS total score during treatment had a significant effect on the IDAP in the patients' samples, therefore suggesting that IDAP does not reflect the temporary depressive state in MDD [130].

The data concluded from this psychiatric research could actually contribute to finding a more specific clinical basis in the differential indication of serotonergic versus noradrenergic antidepressants in the case of depression following stroke. To date, very few studies have been conducted on this issue by means of cortical-evoked potentials. The aforementioned study from Rocco et al. [70] was designed to also correlate the presence of depressive symptoms after stroke with a central serotonergic tone in IDAP; it is reported that the most frequent DS were sadness, lack of energy, loss of interest, sleep difficulties, anhedonia and loss of appetite. It is intriguing that those DS were related to a strong increase in IDAP value, which means a lower serotonergic tone; in fact, this is in line with a similar psychiatric study, which showed a positive correlation of IDAP with the degree of certain somatic symptoms of depression (not PSD), such as loss of appetite and weight, insomnia, and sexual dysfunction. Thus, there is a need for further research aimed at confirming these data, even for depression following stroke.

In this view, the same consideration should be made concerning anger and aggressive behaviour after stroke. Paradoxically, despite the fact that serotonin is thought to play the main role in its pathophysiology, very few studies have been carried out on this issue by means of cortical-evoked potentials.

The reason for better analysing the relationship between aggressive behaviour after stroke and central serotonergic tone is that recent biochemical and neurophysiological evidence suggests that the so-called "serotonergic hypothesis of aggression," which considers the central serotonergic tone as inversely related to aggression, seems somehow obsolete. More specifically, whether 5HT has the same role both in trait anger and in a state condition of anger is still debated.

A recent study [71] showed, by means of IDAP, that an early decrease in serotonergic tone was related to stroke patients who showed a trait anger (i.e., high aggressive behaviour without any change in the post-stroke condition). Otherwise, in sharp contrast to the classical serotonin hypothesis of aggression, the stroke patients who showed a change in post-stroke aggressiveness in spite of its degree (i.e., a state condition of anger), were described to have a significantly lower IDAP value, which indicates a higher central serotonergic tone.

Hence, this datum could also open new possibilities in the treatment strategies of aggressive behaviour after stroke, suggesting that antidepressant medications – especially acting on a serotonergic system – might be the appropriate treatment for stroke patients showing anger or aggressive behaviour, regardless of eventual changes in pre-stroke and current (post-stroke) status. Despite that several studies have shown a good response to SSRI, future researches may contribute to better understand if there is a subgroup of stroke patients with aggressive behaviour who would benefit from a treatment acting on monoaminergic neural pathways other than serotonergic ones.

REFERENCES

[1] Sutton S, Braren M, Zubin J, John ER: Evoked potential correlates of stimulus uncertainty. *Science* 1965;150:1187-1188.

[2] Diner BC, Holcomb PJ, Dykman RA: P300 in major depressive disorder. Psychiatry Res 1985;15:175-184.

[3] Gummow LJ, Dustman RE, Keaney RP: Cerebrovascular accident alters p300 event-related potential characteristics. *Electroencephalogr Clin Neurophysiol* 1986;63:128-137.

[4] Hirata K, Katayama S, Yamazaki K, Fujikane M, Katayama K: Electric field distribution of event-related potentials in stroke patients. *Brain Topogr* 1996;8:279-284.

[5] Evers S, Koch HG, Grotemeyer KH, Lange B, Deufel T, Ringelstein EB: Features, symptoms, and neurophysiological findings in stroke associated with hyperhomocysteinemia. *Arch Neurol* 1997;54:1276-1282.

[6] Tachibana H, Toda K, Sugita M: Event-related potentials in patients with multiple lacunar infarcts. *Gerontology* 1992;38:322-329.

[7] Korpelainen JT, Kauhanen ML, Tolonen U, Brusin E, Mononen H, Hiltunen P, Sotaniemi KA, Suominen K, Myllylä VV: Auditory p300 event related potential in minor ischemic stroke. *Acta Neurol Scand* 2000;101:202-208.

[8] Knight RT, Scabini D, Woods DL, Clayworth CC: Contributions of temporal-parietal junction to the human auditory p3. *Brain Res* 1989;502:109-116.

[9] Buchsbaum M, Silverman J: Stimulus intensity control and the cortical evoked response. *Psychosom Med* 1968;30:12-22.

[10] Scherg M, Von Cramon D: Two bilateral sources of the late aep as identified by a spatio-temporal dipole model. *Electroencephalogr Clin Neurophysiol* 1985;62:32-44.

[11] Norra C, Mrazek M, Tuchtenhagen F, Gobbelé R, Buchner H, Sass H, Herpertz SC: Enhanced intensity dependence as a marker of low serotonergic neurotransmission in borderline personality disorder. *J Psychiatr Res* 2003;37:23-33.

[12] Carrillo-de-la-Peña MT: One-year test-retest reliability of auditory-evoked potentials (aeps) to tones of increasing intensity. *Psychophysiology* 2001;38:417-424.

[13] Hegerl U, Juckel G: Intensity dependence of auditory-evoked potentials as an indicator of central serotonergic neurotransmission: A new hypothesis. *Biol Psychiatry* 1993;33:173-187.

[14] van Donkelaar EL, Blokland A, Ferrington L, Kelly PA, Steinbusch HW, Prickaerts J: Mechanism of acute tryptophan depletion: Is it only serotonin? *Mol Psychiatry* 2011;16:695-713.

[15] von Knorring L, Perris C: Biochemistry of augmenting/reducing response in visual-evoked potentials. *Neurophysiology* 1981;7:1-8.

[16] Strobel A, Debener S, Schmidt D, Hünnerkopf R, Lesch KP, Brocke B: Allelic variation in serotonin transporter function associated with the intensity dependence of the auditory evoked potential. *Am J Med Genet B Neuropsychiatr Genet* 2003;118:41-47.

[17] Park YM, Lee SH, Lee HJ, Kang SG, Min JA, Chae JH: Association between bdnf gene polymorphisms and serotonergic activity using loudness dependence of auditory-evoked potentials in healthy subjects. *PLoS One* 2013;8:e60340.

[18] Rankin CH, Abrams T, Barry RJ, Bhatnagar S, Clayton DF, Colombo J, Coppola G, Geyer MA, Glanzman DL, Marsland S, McSweeney FK, Wilson DA, Wu CF, Thompson RF: Habituation revisited: An updated and revised description of the behavioural characteristics of habituation. *Neurobiol Learn Mem* 2009;92:135-138.

[19] Kandel ER: An introduction to behavioural neurobiology. In: The cellular basis of behaviour WH Freeman; San Francisco 1975.

[20] Kandel ER, Tauc L: Heterosynaptic facilitation in neurones of the abdominal ganglion of aplysia depilans. *J Physiol* 1965;181:1-27.

[21] Rankin CH, Wicks SR: Mutations of the caenorhabditis elegans brain-specific inorganic phosphate transporter eat-4 affect habituation of the tap-withdrawal response without affecting the response itself. *J Neurosci* 2000;20:4337-4344.

[22] Thompson RF: Habituation: A history. *Neurobiol Learn Mem* 2009;92: 127-134.

[23] Barry RJ: Habituation of the orienting reflex and the development of preliminary process theory. *Neurobiol Learn Mem* 2009;92:235-242.

[24] Thompson RF, Spencer WA: Habituation: A model phenomenon for the study of neuronal substrates of behaviour. *Psychol Rev* 1966;73:16-43.

[25] Groves PM, Thompson RF: Habituation: A dual-process theory. *Psychol Rev* 1970;77:419-450.

[26] Glanzman DL, Thompson RF: Evidence against conduction failure as the mechanism underlying monosynaptic habituation in frog spinal cord. *Brain Res* 1979;174:329-332.

[27] Ezzeddine Y, Glanzman DL: Prolonged habituation of the gill-withdrawal reflex in aplysia depends on protein synthesis, protein phosphatase activity, and postsynaptic glutamate receptors. *J Neurosci* 2003;23:9585-9594.

[28] Rose JK, Rankin CH: Analyses of habituation in caenorhabditis elegans. *Learn Mem* 2001;8:63-69.

[29] Li Q, Roberts AC, Glanzman DL: Synaptic facilitation and behavioural dishabituation in aplysia: Dependence on release of ca2+ from postsynaptic intracellular stores, postsynaptic exocytosis, and modulation of postsynaptic ampa receptor efficacy. *J Neurosci* 2005;25:5623-5637.

[30] Coppola G, Di Lorenzo C, Schoenen J, Pierelli F: Habituation and sensitization in primary headaches. *J Headache Pain* 2013;14.

[31] Brighina F, Palermo A, Fierro B: Cortical inhibition and habituation to evoked potentials: Relevance for pathophysiology of migraine. *J Headache Pain* 2009;10:77-84.

[32] Magis D, Vigano A, Sava S, D'Elia TS, Schoenen J, Coppola G: Pearls and pitfalls: Electrophysiology for primary headaches. *Cephalalgia* 2013;33:526-539.

[33] Coppola G, Pierelli F, Schoenen J: Is the cerebral cortex hyperexcitable or hyperresponsive in migraine? *Cephalalgia* 2007;27:1427-1439.

[34] Magis D, Lisicki M, Coppola G: Highlights in migraine electrophysiology: Are controversies just reflecting disease heterogeneity? *Curr Opin Neurol* 2016;29:320-330.

[35] Katsarava Z, Giffin N, Diener HC, Kaube H: Abnormal habituation of 'nociceptive' blink reflex in migraine--evidence for increased excitability of trigeminal nociception. *Cephalalgia* 2003:814-819.

[36] De Marinis M, Pujia A, Natale L, D'arcangelo E, Accornero N: Decreased habituation of the r2 component of the blink reflex in migraine patients. *Clin Neurophysiol* 2003;114:889-893.

[37] Di Clemente L, Puledda F, Biasiotta A, Viganò A, Vicenzini E, Truini A, Cruccu G, Di Piero V: Topiramate modulates habituation in migraine: Evidences from nociceptive responses elicited by laser evoked potentials. *J Headache Pain* 2013;14.

[38] de Tommaso M, Libro G, Guido M, Losito L, Lamberti P, Livrea P: Habituation of single co2 laser-evoked responses during interictal phase of migraine. *J Headache Pain* 2005;6:195-198.

[39] Valeriani M, de Tommaso M, Restuccia D, Le Pera D, Guido M, Iannetti GD, Libro G, Truini A, Di Trapani G, Puca F, Tonali P, Cruccu G: Reduced habituation to experimental pain in migraine patients: A co(2) laser evoked potential study. *Pain* 2003;105:57-64.

[40] de Tommaso M, Valeriani M, Sardaro M, Serpino C, Fruscolo OD, Vecchio E, Cerbo R, P. L: Pain perception and laser evoked potentials during menstrual cycle in migraine. *J Headache Pain* 2009;10:423-429.

[41] Aurora SK, Cao Y, Bowyer SM, Welch KM: The occipital cortex is hyperexcitable in migraine: Experimental evidence. *Headache* 1999;39:469-476.

[42] Afra J, Mascia A, Gérard P, Maertens de Noordhout A, Schoenen J: Interictal cortical excitability in migraine: A study using transcranial magnetic stimulation of motor and visual cortices. *Ann Neurol* 1998;44:209-215.

[43] Bohotin V, Fumal A, Vandenheede M, Bohotin C, Schoenen J: Excitability of visual v1-v2 and motor cortices to single transcranial magnetic stimuli in migraine: A reappraisal using a figure-of-eight coil. *Cephalalgia* 2003;23:264-270.

[44] Bohotin V, Fumal A, Vandenheede M, Gérard P, Bohotin C, Maertens de Noordhout A, Schoenen J: Effects of repetitive transcranial magnetic stimulation on visual-evoked potentials in migraine. *Brain* 2002;125:912-922.

[45] Mulleners WM, Chronicle EP, Palmer JE, Koehler PJ, Vredeveld JW: Suppression of perception in migraine: Evidence for reduced inhibition in the visual cortex. *Neurology* 2001;56:178-183.

[46] Schoenen J: Deficient habituation of evoked cortical potentials in migraine: A link between brain biology, behaviour and trigeminovascular activation? *Biomed Pharmacother* 1996;50:71-78.

[47] Coppola G, Di Clemente L, Fumal A, Magis D, De Pasqua V, Pierelli F, Schoenen J: Inhibition of the nociceptive r2 blink reflex after supraorbital or index finger stimulation is normal in migraine without aura between attacks. *Cephalalgia* 2007;27:803-808.

[48] Knott JR, Irwin DA: Anxiety, stress, and the contingent negative variation. *Arch Gen Psychiatry* 1973;29:538-541.

[49] Coppola G, Ambrosini A, Di Clemente L, Magis D, Fumal A, Gérard P, Pierelli F, Schoenen J: Interictal abnormalities of gamma band activity in visual evoked responses in migraine: An indication of thalamocortical dysrhythmia? *Cephalalgia* 2007;27:1360-1367.

[50] Chen WT, Wang SJ, Fuh JL, Lin CP, Ko YC, Lin YY: Persistent ictal-like visual cortical excitability in chronic migraine. *Pain* 2011;152:254-258.

[51] Ambrosini A, Rossi P, De Pasqua V, Pierelli F, Schoenen J: Lack of habituation causes high intensity dependence of auditory evoked cortical potentials in migraine. *Brain* 2003;126:2009-2015.

[52] Curio G: Linking 600-hz "spikelike" eeg/meg wavelets ("sigma-bursts") to cellular substrates: Concepts and caveats. *J Clin Neurophysiol* 2000;17:377-396.

[53] Restuccia D, Del Piero I, Martucci L, Zanini S: High-frequency oscillations after median-nerve stimulation do not undergo habituation: A new insight on their functional meaning? *Clin Neurophysiol* 2011;122: 148-152.

[54] Restuccia D, Vollono C, Del Piero I, Martucci L, Zanini S: Somatosensory high frequency oscillations reflect clinical fluctuations in migraine. *Clin Neurophysiol* 2012;123:2050-2056.

[55] Coppola G, Vandenheede M, Di Clemente L, Ambrosini A, Fumal A, De Pasqua V, Schoenen J: Somatosensory evoked high-frequency oscillations reflecting thalamo-cortical activity are decreased in migraine patients between attacks. *Brain* 2005;128:98-103.

[56] Kropp P, Gerber WD: Prediction of migraine attacks using a slow cortical potential, the contingent negative variation. *Neurosci Lett* 1998;257:73-76.

[57] Siniatchkin M, Gerber WD, Kropp P, Voznesenskaya T, Vein AM: Are the periodic changes of neurophysiological parameters during the pain-free interval in migraine related to abnormal orienting activity? *Cephalalgia* 2000;20:20-29.

[58] Coppola G, Pierelli F, Schoenen J: Habituation and migraine. Neurobiol *Learn Mem* 2009;92:249-259.

[59] Coppola G, Crémers J, Gérard P, Pierelli F, Schoenen J: Effects of light deprivation on visual-evoked potentials in migraine without aura. *BMC Neurol* 2011;11.

[60] Schoenen J: Is chronic migraine a never-ending migraine attack? *Pain* 2011;152:239-240.

[61] Chen WT, Wang SJ, Fuh JL, Ko YC, Lee YC, Hämäläinen MS, Lin YY: Visual cortex excitability and plasticity associated with remission from chronic to episodic migraine. *Cephalalgia* 2012;32:537-543.

[62] Cosentino G, Fierro B, Vigneri S, Talamanca S, Paladino P, Baschi R, Indovino S, Maccora S, Valentino F, Fileccia E, Giglia G, Brighina F: Cyclical changes of cortical excitability and metaplasticity in migraine: Evidence from a repetitive transcranial magnetic stimulation study. *Pain* 2014;155:1070-1078.

[63] Pierelli F, Iacovelli E, Bracaglia M, Serrao M, Coppola G: Abnormal sensorimotor plasticity in migraine without aura patients. *Pain* 2013;154:1738-1742.

[64] Coppola G, Di Renzo A, Tinelli E, Iacovelli E, Lepre C, Di Lorenzo C, Di Lorenzo G, Di Lenola D, Parisi V, Serrao M, Pauri F, Fiermonte G, Bianco F, Pierelli F: Evidence for brain morphometric changes during the migraine cycle: A magnetic resonance-based morphometry study. *Cephalalgia* 2015;35:783-791.

[65] Schwedt TJ, Zuniga L, Chong CD: Low heat pain thresholds in migraineurs between attacks. *Cephalalgia* 2015;35:593-599.

[66] Coppola G, Di Renzo A, Tinelli E, Lepre C, Di Lorenzo C, Di Lorenzo G, Scapeccia M, Parisi V, Serrao M, Colonnese C, Schoenen J, Pierelli F: Thalamo-cortical network activity between migraine attacks: Insights from mri-based microstructural and functional resting-state network correlation analysis. *J Headache Pain* 2016;17:100.

[67] Onofrj M, Curatola L, Malatesta G, Colamartino P, Bazzano S, Fulgente T, Ferracci F: Delayed p3 event-related potentials (erps) in thalamic hemorrhage. *Electroencephalogr Clin Neurophysiol* 1992;83:52-61.

[68] Pariente J, Loubinoux I, Carel C, Albucher JF, Leger A, Manelfe C, Rascol O, Chollet F: Fluoxetine modulates motor performance and cerebral activation of patients recovering from stroke. *Ann Neurol* 2001;50:718-729.

[69] Acler M, Robol E, Fiaschi A, Manganotti P: A double blind placebo rct to investigate the effects of serotonergic modulation on brain excitability and motor recovery in stroke patients. *J Neurol* 2009;256:1152-1158.

[70] Rocco A, Afra J, Toscano M, Sirimarco G, Di Clemente L, Altieri M, Lenzi GL, Di Piero V: Acute subcortical stroke and early serotonergic modification: A idap study. *Eur J Neurol* 2007;14:1378-1382.

[71] Toscano M, Viganò A, Puledda F, Verzina A, Rocco A, Lenzi GL, Di Piero V: Serotonergic correlation with anger and aggressive behaviour in acute stroke patients: An intensity dependence of auditory-evoked potentials (idap) study. *Eur Neurol* 2014;72:186-192.

[72] Gainotti G: Emotional behaviour and hemispheric side of the lesion. *Cortex* 1972;8:41-55.

[73] Douven E SS, Verhey FR, van Oostenbrugge RJ, Aalten P, Staals J, Köhler S.: The cognition and affect after stroke - a prospective evaluation of risks (casper) study: Rationale and design. *BMC Neurol* 2016;16.

[74] Nys GM, van Zandvoort MJ, van der Worp HB, de Haan EH, de Kort PL, Jansen BP, Kappelle LJ: Early cognitive impairment predicts long-term depressive symptoms and quality of life after stroke. *J Neurol Sci* 2006;247:149-156.

[75] Towfighi A, Ovbiagele B, El Husseini N, Hackett ML, Jorge RE, Kissela BM, Mitchell PH, Skolarus LE, Whooley MA, Williams LS: Poststroke depression: A scientific statement for healthcare professionals from the american heart association/american stroke association. *Stroke* 2016.

[76] Snaphaan L, van der Werf S, Kanselaar K, de Leeuw FE: Post-stroke depressive symptoms are associated with post-stroke characteristics. *Cerebrovasc Dis* 2009;28:551-557.

[77] Altieri M, Maestrini I, Mercurio A, Troisi P, Sgarlata E, Rea V, Di Piero V, Lenzi GL: Depression after minor stroke: Prevalence and predictors. *Eur J Neurol* 2012;19:517-521.

[78] Kim JS: Post-stroke mood and emotional disturbances: Pharmacological therapy based on mechanisms. *J Stroke* 2016;18:244-255.

[79] Hackett ML, Pickles K: Part i: Frequency of depression after stroke: An updated systematic review and meta-analysis of observational studies. *Int J Stroke* 2014;9:1017-1025.

[80] Tooth LR, Ottenbacher KJ, Smith PM, Illig SB, Linn RT, Gonzales VA, Granger CV: Effect of functional gain on satisfaction with medical rehabilitation after stroke. *Am J Phys Med Rehabil* 2003;82:692-699.

[81] Carson AJ, MacHale S, Allen K, Lawrie SM, Dennis M, House A, Sharpe M: Depression after stroke and lesion location: A systematic review. *Lancet* 2000;356:122-126.

[82] Wei N, Yong W, Li X, Zhou Y, Deng M, Zhu H, Jin H: Post-stroke depression and lesion location: A systematic review. *J Neurol* 2015;262:81-90.

[83] Folstein MF, Maiberger R, McHugh PR: Mood disorder as a specific complication of stroke. *J Neurol Neurosurg Psychiatry* 1977;40.

[84] Luijendijk HJ, Stricker BH, Wieberdink RG, Koudstaal PJ, Hofman A, Breteler MM, Tiemeier H: Transient ischemic attack and incident depression. *Stroke* 2011;42:1857-1861.

[85] Shi Y, Xiang Y, Yang Y, Zhang N, Wang S, Ungvari GS, Chiu HF, Tang WK, Wang Y, Zhao X, Wang Y, Wang C: Depression after minor stroke: Prevalence and predictors. *J Psychosom Res* 2015;79:143-147.

[86] Starkstein S, Berthier M, Fedoroff P, Price T, Robinson R: Anosognosia and major depression in 2 patients with cerebrovascular lesions. *Neurology* 1990;40:1380-1382.

[87] Kim JM, Stewart R, Kang HJ, Kim SW, Shin IS, Kim HR, Shin MG, Kim JT, Park MS, Cho KH, Yoon JS: A longitudinal study of slc6a4 DNA promoter methylation and poststroke depression. *J Psychiatr Res* 2013;47:1222-1227.

[88] Andersen G, Vestergaard K, Lauritzen L: Effective treatment of poststroke depression with the selective serotonin reuptake inhibitor citalopram. *Stroke* 1994;25

[89] Stamenkovic M, Schindler S, Kasper S: Poststroke depression and fluoxetine. *Am J Psychiatry* 1996;153:446-447.

[90] Paige SR, Fitzpatrick DF, Kline JP, Balogh SE, Hendricks SE: Event-related potential amplitude/intensity slopes predict response to antidepressants. *Neuropsychobiology* 1994;30:197-201.

[91] Vandoolaeghe E, van Hunsel F, Nuyten D, Maes M: Auditory event related potentials in major depression: Prolonged p300 latency and increased p200 amplitude. *J Affect Disord* 1998;48:105-113.

[92] Gallinat J, Bottlender R, Juckel G, Munke-Puchner A, Stotz G, Kuss HJ, Mavrogiorgou P, Hegerl U: The loudness dependency of the auditory evoked n1/p2-component as a predictor of the acute ssri response in depression. *Psychopharmacology* (Berl) 2000;148:404-411.

[93] Linka T, Müller BW, Bender S, Sartory G: The intensity dependence of the auditory evoked n1 component as a predictor of response to citalopram treatment in patients with major depression. *Neurosci Lett* 2004;367:375-378.

[94] Yanai I, Fujikawa T, Osada M, Yamawaki S, Touhouda Y: Changes in auditory p300 in patients with major depression and silent cerebral infarction. *J Affect Disord* 1997;46:263-271.

[95] Montgomery SA, Asberg M: A new depression scale designed to be sensitive to change. 1979; 134: *British Journal of Psychiatry Res* 1979;134:382–389.

[96] Loubinoux I, Kronenberg G, Endres M, Schumann-Bard P, Freret T, Filipkowski RK, Kaczmarek L, Popa-Wagner A: Post-stroke depression: Mechanisms, translation and therapy. *J Cell Mol Med* 2012;16:1961-1969.

[97] Noonan K, Carey LM, Crewther SG: Meta-analyses indicate associations between neuroendocrine activation, deactivation in neurotrophic and neuroimaging markers in depression after stroke. *J Stroke Cerebrovasc Dis* 2013;22:124-135.

[98] Aybek S, Carota A, Ghika-Schmid F, Berney A, Melle GV, Guex P, Bogousslavsky J: Emotional behaviour in acute stroke: The lausanne emotion in stroke study. *Cogn Behav Neurol* 2005;18:37-44.

[99] Santos C, Caeiro L, Ferro JM, Albuquerque R, Luísa Figueira M: Anger, hostility and aggression in the first days of acute stroke. *Eur J Neurol* 2006;13:351-358.

[100] Kim JS, Choi S, Kwon SU, Seo YS: Inability to control anger or aggression after stroke. *Neurology* 2002;58:1106-1108.

[101] Williams JE, Nieto FJ, Sanford CP, Couper DJ, Tyroler HA: The association between trait anger and incident stroke risk: The atherosclerosis risk in communities (aric) study. *Stroke* 2002;33:13-19.

[102] Ghika-Schmid F, van Melle G, Guex P, Bogousslavsky J: Subjective experience and behaviour in acute stroke: The lausanne emotion in acute stroke study. *Neurology* 1999;52:22-28.

[103] Cummings JL, Bogousslavsky J: Emotional consequences of focal brain lesions: An overview. in Bogousslavsky J and Cummings JL (eds) Behaviour and mood disorders in focal brain lesions Cambridge University Press 2000:1-20.

[104] Giacalone E, Tansella M, Valzelli L, Garattini S: Brain serotonin metabolism in isolated aggressive mice. *Biochem Pharmacol* 1968; 17:1315-1327.

[105] Brown GL, Goodwin FK, Ballenger JC, Goyer PF, Major LF: Aggression in humans correlates with cerebrospinal fluid amine metabolites. *Psychiatry Res* 1979;1:131-139.

[106] Linnoila M, Virkkunen M, Scheinin M, Nuutila A, Rimon R, Goodwin FK: Low cerebrospinal fluid 5-hydroxyindoleacetic acid concentration differentiates impulsive from nonimpulsive violent behaviour. *Life Sci* 1983;33:2609-2614.

[107] Davidson RJ, Putnam KM, Larson CL: Dysfunction in the neural circuitry of emotion regulation--a possible prelude to violence. *Science* 2000;289:591-594.
[108] Coccaro EF: Central serotonin and impulsive aggression. *Br J Psychiatry* Suppl 1989;8:52-62.
[109] Moss HB, Yao JK, Panzak GL: Serotonergic responsivity and behavioural dimensions in antisocial personality disorder with substance abuse. *Biol Psychiatry* 1990;28:325-338.
[110] Houston RJ, Stanford MS: Mid-latency evoked potentials in self-reported impulsive aggression. *Int J Psychophysiol* 2001;40:1-15.
[111] Rosenstein GZ, Furman V, Sohmer H, Attias J, Abraham F: Single p100 visual-evoked potential analyses in man. *Int J Neurosci* 1994;79:251-265.
[112] Carrillo-de-la-Peña MT: Erp augmenting/reducing and sensation seeking: A critical review. *Int J Psychophysiol* 1992;12:211-220.
[113] de Boer SF, Koolhaas JM: 5-ht1a and 5-ht1b receptor agonists and aggression: A pharmacological challenge of the serotonin deficiency hypothesis. *European Journal of Pharmacology* 2005;526:125-139.
[114] Coccaro EF: Central serotonin and impulsive aggression. *Br J Psychiatry* Suppl 1989:52-62.
[115] van der Vegt BJ, Lieuwes N, Cremers TI, de Boer SF, Koolhaas JM: Cerebrospinal fluid monoamine and metabolite concentrations and aggression in rats. *Horm Behav* 2003;44:199-208.
[116] Evers S, Afra J, Frese A, Goadsby PJ, Linde M, May A, Sándor PS: European federation of neurological societies. Efns guideline on the drug treatment of migraine--revised report of an efns task force. *Eur J Neurol* 2009;16:968-981.
[117] Manack AN, Buse DC, Lipton RB: Chronic migraine: Epidemiology and disease burden. *Curr Pain Headache Rep* 2011;15.
[118] Sándor PS, Afra J, Ambrosini A, Schoenen J: Prophylactic treatment of migraine with beta-blockers and riboflavin: Differential effects on the intensity dependence of auditory evoked cortical potentials. *Headache* 2000;40:30-35.
[119] Shirahige L, Melo L, Nogueira F, Rocha S, Monte-Silva K: Efficacy of noninvasive brain stimulation on pain control in migraine patients: A systematic review and meta-analysis. *Headache* 2016;56:1565-1596.

[120] Viganò A, D'Elia TS, Sava SL, Auvé M, De Pasqua V, Colosimo A, Di Piero V, Schoenen J, Magis D: Transcranial direct current stimulation (tdcs) of the visual cortex: A proof-of-concept study based on interictal electrophysiological abnormalities in migraine. *J Headache Pain* 2013;14.

[121] Buonfiglio M, Toscano M, Puledda F, Avanzini G, Di Clemente L, Di Sabato F, Di Piero V: Lack of habituation of evoked visual potentials in analytic information processing style: Evidence in healthy subjects. *Neurol Sci* 2015;36:391-395.

[122] Di Sabato F, Buonfiglio M, Mandillo S: Analytic information processing style in migraineurs. *Neurol Sci* 2013;34:1145-1150.

[123] Andersen G, Ingeman-Nielsen M, Vestergaard K, Riis JO: Pathoanatomic correlation between poststroke pathological crying and damage to brain areas involved in serotonergic neurotransmission. *Stroke* 1994;25:1050-1052.

[124] Murai T, Barthel H, Berrouschot J, Sorger D, von Cramon DY, Müller U: Neuroimaging of serotonin transporters in post-stroke pathological crying. *Psychiatry Res* 2003;123:207-211.

[125] Choi-Kwon S, Han K, Choi S, M S, Kim YJ, Song H: Poststroke depression and emotional incontinence: Factors related to acute and subacute stages. *Neurology* 2012;78:1130-1137.

[126] Hackett ML, Yang M, Anderson CS, Horrocks JA, House A: Pharmaceutical interventions for emotionalism after stroke. *Cochrane Database Syst Rev* 2010:CD003690.

[127] Rampello L, Chiechio S, Nicoletti G, Alvano A, Vecchio I, Raffaele R, Malaguarnera M: Prediction of the response to citalopram and reboxetine in post-stroke depressed patients. *Psychopharmacology* (Berl) 2004;173:73-78.

[128] Hackett ML, Anderson CS, House A, Xia J: Interventions for treating depression after stroke. *Cochrane Database Syst Rev* 2008:CD003437;4.

[129] Mead GE, Hsieh CF, Lee R, Kutlubaev MA, Claxton A, Hankey GJ, Hackett ML: Selective serotonin reuptake inhibitors (ssris) for stroke recovery. *Cochrane Database Syst Rev* 2012:CD009286 2012.

[130] Linka T, Sartory G, Wiltfang J, Müller BW: Treatment effects of serotonergic and noradrenergic antidepressants on the intensity dependence of auditory erp components in major depression. *Neurosci Lett* 2009;463:26-30.

Biographical Sketches

Dt. Massimiliano Toscano, MD, PhD

Education:

2005	Medical Degree, University of Rome “La Sapienza,” summa cum laude
2013	Residence in neurology, Dept. Neurological Sciences, University of Rome “La Sapienza,” summa cum laude
2016	PhD in Behavioural Neuroscience, Dept. Neurological Sciences, Dept. of Psychology, University of Rome “La Sapienza”

Present Appointments and Previous Experience:

2015–present	Research Assistant, Dept. Neurological Sciences, University of Rome “La Sapienza”
2013–present	Guest Neurologist of the Stroke Unit and the Headache Clinic, Dept. Neurological Sciences, University of Rome “La Sapienza”
2013–present	Neurologist of the rehabilitation clinic A.N.F.F.A.S. Rome onlus.
2013–2014	Lecturer in Clinical Neurology, learning related mechanisms and disability, I.R.S.E.F. Research Institute
2005–2008	Guest doctor Stroke Unit and the Headache Clinic, Dept. Neurological Sciences, University of Rome “La Sapienza”
2005–2007	Lecturer in Clinical Neurology. Graduating Course in Speech Therapy, University of Rome “Tor Vergata”

Research Focus:

Cerebrovascular diseases (pathophysiology of stroke in the acute phase, mechanisms of post-stroke recovery, clinical application of functional neuroimaging techniques, neurosonology, psychiatric comorbidity)

Behavioural neuroscience (behavioural neurophysiology, learning related mechanisms and disability, speech related mechanisms, neurophysiological correlates of psychiatric diseases)

Pain (morphofunctional correlates of pain in humans, pathophysiology of primary headaches, migraine, psychiatric comorbidity)

Other Activities:

Member of the ESNCH (European Society of Neurosonology and Cerebral Hemodynamics) and the SISC (Società Italiana per lo Studio delle Cefalee)

Member of the Executive Committee of the SISS (Società Italiana per lo Studio dello Stroke)

Member of the Executive Committee of the I.R.S.E.F. Research Institute.

Dr. Alessandro Viganò, MD, PhD

Education:

2007 Medical Degree, University of Rome "La Sapienza," with the highest honors

2012 PhD in Biophysics, Dept. Anatomy, Histology, Orthopedics and Forensic Medicine, University of Rome "La Sapienza," with the highest honors

Current Position:

2013 In-training fellow of Neurology in the Neurology Residency, University of Rome "La Sapienza."

Research Focus:

Cerebrovascular diseases (pathophysiology of stroke in the acute phase, mechanisms of post-stroke recovery, clinical application of functional neuroimaging techniques, neurosonology, psychiatric comorbidity)

Behavioural neuroscience (behavioural neurophysiology, learning related mechanisms and disability, speech related mechanisms, neurophysiological correlates of psychiatric diseases)

Pain (morphofunctional correlates of pain in humans, pathophysiology of primary headaches, migraine, psychiatric comorbidity)

Prof. Vittorio Di Piero, MD, PhD

Education:

1979 Medical Degree, University of Rome "La Sapienza," summa cum laude

Present Appointments and Previous Experience:

2000–present Associate Professor of Neurology, Dept. Neurological Sciences, University of Rome "La Sapienza," executive coordinator of the Stroke Unit and the Headache Clinic

1989–2000 Senior Registrar - Dept. Neurological Sciences, University of Rome "La Sapienza"

Visiting Appointments:

1988–1989 Honorary Research Fellow, Royal Postgraduate School, Hammersmith Hospital, London

1984–1985 Research Fellow, Dept. Biomedical Technologies, Istituto S. Raffaele, Milan

Research Focus:

Cerebrovascular diseases (pathophysiology of stroke in the acute phase, mechanisms of post-stroke recovery, clinical application of functional neuroimaging techniques)

Aging and dementia (cognitive decline and its clinical and morphological correlates, perfusional derangements during degenerative and vascular dementia)

Pain (morphofunctional correlates of pain in humans, pathophysiology of primary headaches, migraine, psychiatric comorbidity).

Other Activities:

Member of the European Neurological Society; the International Stroke ociety; the European Stroke Council; the International Association for the Study of Pain; the Società Italiana di Neurologia and the Società Italiana per lo Studio delle Cefalee.

Member and President Elected of the Executive Committee of the Società Italiana per lo Studio dello Stroke, and the Associazione per la promozione delle Ricerche Neurologiche.

In: Evoked Potentials and Electrical Stimulation ISBN: 978-1-53611-075-3
Editor: Timothy Eisenhower

Chapter 3

ELECTRICAL STIMULATION IN STROKE PATIENTS

Trinidad Sentandreu-Maño*
Department of Physiotherapy,
University of Valencia, Valencia, Spain

ABSTRACT

Surface electrical stimulation is one of the tools suggested for the rehabilitation treatment of patients following a stroke. Although its use is not very widespread in current clinical practice, it is regarded as a non-invasive, low-cost, reliable and easy-to-use treatment which could stimulate recovery in stroke patients. In recent years, the understanding of motor learning, neuronal plasticity and functional recovery after brain damage has improved, and the new findings in neuroscience have given a fresh impetus to research in neurorehabilitation. The number of existing studies regarding electrical stimulation and the stroke patient shows the interest which the subject still provokes and the need to provide more evidence that will assure its beneficial effect in clinical practice. Current lines of investigation are principally focussed on the recovery of the upper-limb, lower-limb and gait, hemiplegic shoulder, dysphagia, urinary incontinence and visuo-spatial neglect. Additionally, different modalities of application are presented and the most effective protocols are investigated, including their complementary use with other therapeutic strategies and new technologies. In this chapter, these points are

* Corresponding Author Email: trinidad.sentandreu@uv.es.

extensively discussed to showcase the latest evidence, shed more light on this intervention in the field of neurological rehabilitation and establish future lines of interest in which advances in neuroscience and technological development play a significant role.

Keywords: surface electrical stimulation, stroke, neurorehabilitation

INTRODUCTION

Neuromuscular electrical stimulation (NMES) involves the use of electrical current to estimulate neural and muscular structures (Rodríguez, 2004). Although NMES can be applied using superficial, percutaneous or implanted electrodes (IJzerman et al., 2009), in this chapter we will further explore the superficial electrical stimulation as a therapeutic strategy, presented as a non-invasive (Díaz et al., 2005), low-cost, reliable and easy-to-use treatment (Glanz et al., 1996; Sentandreu et al., 2011). NMES has been applied for different purposes and in different pathologies. Among such pathologies, it has been suggested for the treatment of patients following a stroke (Howlett et al., 2015).

A stroke is a sudden, non-convulsive loss of neurological function due to an ischemic or haemorrhagic intracranial vascular event (Mehrholz et al., 2013). It is the main cause of adult disability and the third most common cause of death in Europe (Scutt et al., 2015). The incidence of stroke increases with age (Rigau et al., 2009) and due to the progressive aging of the population more health care requirements are expected in the case of these patients (Bermejo et al., 2009). In addition to disturbances in the cognitive and affective domains, the loss of motor function represents the most serious condition after a stroke (Soekadar et al., 2015). In this respect, the rehabilitation of patients after a stroke constitutes a worldwide challenge (Liu et al., 2016) and innovative methodologies for the purpose of restorative neurorehabilitation are urgently required in order to reduce long-term disability (Dutta et al., 2014).

The first paper reporting on the use of NMES in neurological patients was published in 1961. Liberson et al. (1961) described the use of electrical current to stimulate the dorsal flexors of the foot in stroke patients to prevent drop-foot during the swing phase of gait. Since then, the understanding of motor learning, neuronal plasticity and functional recovery following brain damage has improved and the new findings in neuroscience have given a new impetus to research in neurorehabilitation. This improved understanding has led to the identification of new clinical roles, challenges and emerging research in the field of NMES and stroke patients.

The vast number of existing studies concerning this subject is indicative of an ongoing interest and an increasing need to provide further evidence that will ensure the beneficial effect thereof in clinical practice. Current lines of investigation are principally focussed on the recovery of the upper-limb, lower-limb and gait, hemiplegic shoulder, dysphagia, urinary incontinence and visuo-spatial neglect; different modalities of application are presented and the most effective protocols are investigated, including their complementary use with other therapeutic strategies and new technologies. These points are discussed in greater detail within the framework of this chapter to show the most recent evidence, shed more light on this intervention in the field of neurological rehabilitation and set out future lines of interest in which advances in neuroscience and technological development play a significant role.

UPPER-LIMB

Impairment of the upper limb is one of the most frequent consequences of stroke (Mangold et al., 2009) and directly impacts on the patient's function and quality of life. It is estimated that less than 50% of stroke survivors will recover arm function (Nakayama et al., 1994), which will exert a great economic, social and personal toll. Among the different therapeutic tools, NMES has been proposed in several studies as a strategy in the rehabilitation of the upper limb of stroke patients, and different reviews and meta-analyses support its use (Howlett et al., 2015; Nascimento et al., 2014). However, despite of it, this topic remains at present a research focus for various authors as can be seen with the number of studies in the literature.

The most recent research on electrical stimulation and upper limb recovery focuses mainly on the most effective protocols by comparing different variables of the applied programmes (i.e., NMES modalities, stimulation parameters, electrode placement, etc.), the characteristics of the patients who can benefit most from this treatment, its effectiveness as a complement to other emerging therapies, and its effects not only locally but also its impact on the central nervous system.

Regarding the main NMES modalities found in the randomised controlled trials (RCTs) reviewed, certain studies have applied sensory or motor electrical stimulation thresholds. In the first approach, the stimulation evokes a sensory reaction without muscle contraction. Some authors also include the transcutaneous electrical nerve stimulation (TENS) modality in this group. In the second approach, where a muscular contraction is provoked, several applications can be distinguished: cyclic NMES (without voluntary activity by the patient), sensor or electromyographic (EMG)-triggered NMES (activated voluntarily by the patient) and functional electrical stimulation (FES) (associated with the performance of functional tasks). In some of these cases, NMES treatment was applied by orthosis. One of the most used models in the studies reviewed is the NESS H200 surface FES system for upper limbs (Bioness Inc., Santa Clarita, CA, USA). This model combines wrist-hand orthosis with muscle activation of the paralyzed forearm and hand via integrated surface electrodes. It allows consistent replication of the grasp, hold and release functions of the hand.

Few previous studies have directly compared electrical stimulation modalities, and although many modalities have been established, conclusive proof of their effects is still lacking (Hara et al., 2008). De Kroon et al. (2008) and Boyaci et al. (2013) compared cyclic NMES and EMG-triggered NMES, and found no difference in outcomes. These findings disagree with the review by Kroon et al. (2005), which argued that electrical stimulation triggered by voluntary movement appeared to be more beneficial than other NMES modes. It has been hypothesized that the intention to execute a movement, along with peripheral activity, reinforces plastic changes further compared to cyclic stimulation (Quandt and Hummel, 2014). A recently published study (Wilson et al., 2016) that compared the effect of cyclic NMES, EMG-triggered NMES and sensory stimulation, concluded that all groups exhibited significant improvement of impairment and functional limitation in the upper limb, though there was no difference based on the type of electrical stimulation administered.

Electrogoniometers, bend sensors, touch sensitive mats, and accelerometers are among the external sensors that have been incorporated into NMES systems (Knutson et al., 2015). Researchers also continue to explore the use of signal collection from the impaired upper limb to not merely trigger the onset of a pre-set intensity and duration of stimulation but also to control the intensity and timing of stimulation. A challenge for EMG-triggered NMES modalities is that the effort required from patients to contract the muscle does not develop flexor synergies or hypertonia (Knutson et al., 2015).

Contralaterally controlled functional electrical stimulation (CCFES) is an innovative method of sensor-controlled stimulation that uses movement from the unimpaired side to control the timing and intensity of stimulation to the paretic side. The hand system consists of a glove with bend sensors worn on the non-paretic hand and a multi-channel stimulator that delivers stimulation to the paretic hand extensors with an intensity that is proportional to the degree of opening of the glove. A recent study indicated that CCFES improved hand dexterity more than cyclic NMES in chronic stroke patients (Knutson et al., 2016).

Recent advances in neurotechnology have led to the development of brain-computer interfaces (BCIs) that translate brain activity into control signals of computers (Venkatakrishnan et al., 2014). This technology provides a means to decode mental states and activate devices according to user intentions, and could provide a direct feedback on the engagement of motor areas of the brain surrounding the lesion site (Millán et al., 2010). BCI combined with output devices such as FES has also been studied (Silvoni et al., 2011). BCI with a FES system has been explored as a potential rehabilitation tool for motor learning after stroke because FES can assist such patients in learning and practicing the desired coordinated movements (Müller-Putz et al., 2005). Biasiucci et al. (2013) reported preliminary results of a BCI with a FES system for stroke rehabilitation where the user's intention to perform an extension movement of the affected hand was detected through a BCI and used to activate a FES device. These results confirmed the beneficial effects of direct muscle stimulation according to user intentions to perform a motor task. Along the same line, Daly et al. (2009) carried out a case study which demonstrated the feasibility of combining BCI and FES for motor learning in stroke patients. After training sessions, volitional control of isolated finger extension was obtained. These results showed that combining BCI and FES promoted motor recovery. The authors argued that this improvement could be attributed to BCI, FES, specific target-oriented physical practice or a combination thereof. In this way, RCTs are required to determine whether BCI actually can have an additive effect beyond FES alone and to recognise the characteristics of stroke patients who can successfully use BCI as a motor learning paradigm. On the other hand, Kim et al. (2016) investigated the effects of action observational training (AOT) plus BCI-based FES on improving upper arm function in stroke patients. Their findings suggest that AOT plus BCI-based FES has a positive influence on upper extremity performance by improving functional mobility and ROM of paretic stroke patients.

Specific stimulation parameters are the subject of debate (Quandt and Hummel, 2014). As is also pointed out in other reviews, the comparison of stimulation parameters is complicated due to the disparity of variables or the omission of data. It has been mentioned that neurophysiological responses vary according to the parameters used, and though their clinical relevance has not been verified, even they may not be crucial in determining the effect of electrical stimulation (de Kroon et al., 2005). The most commonly used were low frequency rectangular biphasic currents, with frequency values of 20 and 100 Hz. The intensity was adapted to the sensory or motor threshold, the pulse duration varied between 100 and 500 μs, and duty cycles as well as ramping up/down periods presented different combinations, in addition to treatment doses. Cauraugh et al. (2003) compared protocols with different times of contraction, concluding that better results were obtained with stimuli of 10 s. Doucet and Griffin (2013) compared protocols with different frequencies and concluded that the effects differ depending on the frequency applied. Sentandreu et al. are currently finalizing a study where two experimental groups with different stimulus frequencies and a control group with conventional therapy have been compared. Their results will be published soon. Clair-Auger et al. (2012) investigated the effect of pulse duration on the paretic and non-paretic arm and found a significant increase in elbow flexion torque for the former when using a wide-pulse stimulation scheme (1 ms). It was speculated that wide-pulse stimulation may lead to a greater reflexive recruitment in the paretic arm, highlighting that pathological changes influence stimulation response. Conforto et al. (2010) concluded that sessions of repetitive peripheral nerve sensory stimulation (PSS) could facilitate motor function after subacute stroke depending on the intensity of the stimulation used. The authors proposed that careful dose-response studies are needed to optimise parameters of PSS before designing costly, larger, double-blind, multi-centre clinical trials.

Regarding treatment doses, in the meta-analysis of Pomeroy et al. (2006), it was pointed out that the length of time that NMES is applied could influence its effectiveness. Some studies have also remarked that one of the reasons for the lack of effectiveness of some programmes could be due to the short duration of treatment (Rosewilliam et al., 2012; Popovic et al., 2002), as well as the need for longer protocols in patients with severe impairments (Mangold et al., 2009; Alon et al., 2008; Popovic et al., 2003). Page et al. (2012) compared the efficacy of different rehabilitation protocols in subacute stroke patients and found that 120 minutes of repetitive task-specific practice combined with electrical stimulation was superior compared to shorter durations of stimulation. In contrast, Hsu et al. (2010) argued that higher and lower doses of NMES led to similar improvements in motor function. A minimum of 10 hours of NMES in combination with regular rehabilitation may improve recovery of arm function in acute stroke patients with severe motor deficit. Along the same line, Kowalczewski et al. (2007) compared two protocols of high and low intensity and only some measures obtained significant changes. Lourenção et al. (2005) reported that six months of FES were necessary for a significant improvement in grip speed, and it was hypothesized that five months would be necessary for an effective programme to improve upper limb function.

As for the electrode placement and stimulated musculature, there is also a diversity of options. Some studies have applied NMES at a unique segmental level in agonist and/or antagonist musculature, and others have made a multi-segmental application, applying electrical stimulation in several segments of the upper limb. There is evidence that greater effect sizes are associated with movements that stimulate functionally useful movements, as seen in studies that stimulate more than one joint (Farmer et al., 2014) such as that by Meadmore et al. (2014). They applied FES to three muscle groups in the upper limb and completed coordinated, repetitive, goal-oriented movements using real objects, controlling the FES signal to each muscle using advanced iterative learning control algorithms in order to ensure that the resulting movement precisely coincided with unimpaired task competition. The positive results indicated that multiple muscle stimulation along with the application of iterative learning technology is promising in stroke rehabilitation and may prove effective in reducing upper limb impairments. De Kroon et al. (2004) compared the stimulation of extensors and flexor-extensors and found no difference. Hara et al. (2007) argued that the combined use of distal EMG-NMES and proximal stimulation (anterior deltoid and triceps brachii) could be more efficient on a functional level.

Concerning patient characteristics, some studies have pointed out the importance of suitable sample selection (Powell et al., 1999; Sentandreu et al., 2011). It has also been claimed that motivated patients (Kroon et al., 2005) with moderate impairment (Powell et al., 1999) and residual motor function (Popovic et al., 2003; Powell et al., 1999) could benefit from NMES. According to this, Chae et al. (2008) added that mild to moderately impaired subjects could benefit most.

Recent studies have investigated NMES combined with other emerging physical therapy approaches and new technologies such as mirror therapy, mental imagery training, bilateral movement training, constraint-induced movement therapy (Page et al., 2006) and robotics. Superior effects were shown in studies employing protocols where electrical stimulation was combined with the mirror therapy (Kim et al., 2015; Lin et al., 2014; Yun et al., 2011). Hong et al. (2012) combined EMG-triggered FES with mental imagery training and observed an advantage over FES alone. In contrast, Hemmen et al. (2007) investigated the effect of movement imagery combined with EMG-triggered NMES compared with a conventional NMES programme and concluded that there was no difference between the two therapeutic protocols. With regard to bilateral movement training, on the one hand, Cauraugh et al. (2002, 2005) concluded that the results where more prevalent in groups where EMG-triggered NMES was applied along with bilateral movement training. Knutson et al. (2007) as we have described before, also obtained more benefits in CCFES compared with cyclic NMES. This modality also combines the effects of NMES, active and goal-oriented training, and bimanual symmetric exercises involving the unaffected upper limb in the training of the affected upper limb. On the other hand, Singer et al. (2013) did not show that a protocol applying bilateral tasks plus EMG-triggered NMES was superior to the unilateral equivalent.

With regard to robotic systems, there is a paucity of information regarding the comparative benefit of NMES (McCabe et al., 2015). O'Connor et al. (2015) investigated the feasibility of using the iPAM Mk2 dual robot rehabilitation system with NMES to provide simultaneous hand opening exercises during assisted reaching exercises for stroke survivors who have moderate to severe arm impairment. Based on the promising findings in this report, the authors suggested that future work be planned to explore the possibilities of further combinations, including in lower limb rehabilitation, and to optimise the characteristics of robotic and stimulation parameters. Hu et al. (2015) concluded that NMES robot-assisted wrist training was more effective than the pure robot training. The additional NMES application in the treatment could result in more improvements in distal motor functions and faster rehabilitation progress. McCabe et al. (2015) compared the response to upper-limb treatment using robotics plus motor learning (ML), FES plus ML and ML alone, in a group of chronic severely impaired stroke survivors. The authors did not find any difference in treatment response across the three intervention groups according to the measures of joint movement coordination or complex functional task performance. On the other hand, Daly et al. (2005) pointed out that comparison of NMES and robotics resulted in different outcomes depending on the therapy applied.

Despite the increased knowledge regarding neural plasticity and brain mapping studies, the specific mechanisms of action of these different electrical stimulation modalities are not clear. It has been speculated that therapeutic effects are probably caused by a combination of local and central effects. At the local level, reference has been made to changes in muscular strength, modification of viscoelastic characteristics and increase of blood flow, among others (de Kroon et al., 2004; Ring et al., 2005). Quandt and Hummel (2014) argued that NMES could influence cortical reorganization. It is hypothesized that either concurrent stimulation of afferent fibres, or antidromic stimulation lead to enhanced synaptic remodelling, but evidence is still lacking. Hara et al. (2008) suggested that sensory components of large afferent fibre activation, propioceptive input, and increased cognitive sensory attention could enhance voluntary movement and function. NMES is associated with concomitant physiologic changes in the brain including activation of primary sensory and motor areas and the supplementary motor area, reduction of intracortical inhibition, and increased amplitude of motor-evoked potentials (Chae et al., 2008). Currently, with access to advanced neuroimaging tests, many studies are aimed at investigating the changes that NMES causes at the level of the central nervous system. Several studies demonstrated evidence of central mechanisms using neurophysiologic experiments such as reaction time and fMRI (Chae et al., 2008). Kimberley et al. (2004) indicated that the NMES programme resulted in improvements in the functionality of the hand and these were associated with an increase in cortical intensity in the ipsilateral primary sensory cortex. According to these authors, Quandt and Hummel (2014) argued that most of the reviewed studies found changes of neural activation after NMES use and a trend towards severe impairment leading to activation of the contralesional sensorimotor cortex can be observed. On the other hand, less impaired patients tend to recruit the ipsilesional site. The literature of cortical reorganization during stroke recovery, describing the relationship of lateralization and functional outcome, supports these conclusions.

Lower-Limb and Gait

Post-stroke gait dysfunction is among the most investigated neurological gait disorders and one of the principal goals of post-stroke rehabilitation (Lindquist et al., 2007). At the level of the lower limb, paresis, along with the inability to grade muscle contractions, poor motor coordination, poor endurance, spasticity, and impaired balance have significant consequences on ambulation (Eng et al., 2007). In particular, the inability to dorsiflex the ankle during the swing phase of gait affects approximately 30% of patients following a stroke, which increases the risk of falls and mortality (Bosch et al., 2014). These impairments have an important impact in patient's life and entail considerable costs for health and social services (Evers et al., 2004).

With reference to the electrical stimulation modalities, NMES to the lower extremities is mainly used by post-stroke patients as an alternative to mechanical orthotic devices during gait or as a training modality during rehabilitation (Kafri et al., 2015). In the second approach, the same modalities described in the upper limb may be observed. Nevertheless, the most widely used alternative is FES, which is associated with the performance of functional activities.

In 1961, Liberson et al. (1961) used FES as a substitute for traditional drop-foot correction within a hemiplegic population. By triggering a foot switch, a single-channel electrical stimulation device stimulated the common fibular nerve via a surface electrode, resulting in dorsiflexion during the swing phase of gait. This led to the first commercially available surface electrode stimulator. Since then, advances in technology and the development of wireless FES systems for the purpose of drop-foot correction have become commercially available and are used as an alternative to the ankle-foot orthosis (AFO) (Kafri et al., 2015). A trigger is used to stimulate the muscles during the specific phases of gait. The most common trigger is a foot switch, which is activated by the user's weight upon initial contact. The dorsiflexion can also be stimulated via the use of a tibial tilt sensor capable of detecting the tibial position during the terminal stance (Bosch et al., 2014). Recently, several devices have been commercially introduced for drop-foot correction. Relevant examples within this context would be the WalkAid® (Hanger Orthopaedic group), a surface peroneal nerve stimulator controlled by a tilt sensor attached just below the knee; the NESS L300™, a new-generation surface electrical stimulation system for drop-foot controlled by a heel switch that wirelessly communicates with the stimulator integrated within a semi-rigid orthosis positioned just below the knee; and the Odstock Dropped Foot Stimulator, which represents one of the most widely used and best documented surface stimulation systems. A portable gait event detection device is essential for ensuring the appropriate timing of the stimulation. A variety of sensors, such as goniometers, gyroscopes, accelerometers, inclinometers or force sensitive resistors and various sensor combinations have been developed in order to meet these requirements (Kafri et al., 2015).

As was described in the upper limb section, CCFES has also been used to improve recovery of ankle dorsiflexion in chronic and subacute stroke patients. Initial studies provided preliminary evidence in support of the efficacy thereof (Knutson et al., 2012).

There has been a growing interest in employing BCI technology in combination with FES in post-stroke gait therapy (McCrimmon et al., 2015). Chung et al. (2015) conducted a pilot study in order to determine the BCI-FES effects on balance and gait function in patients with stroke. The results of this study suggested that BCI-FES training constituted a more effective exercise for the amelioration of balance and gait function as opposed to FES training alone in patients with stroke. The authors argued that further studies are needed to investigate its long-term intervention effects. In addition, McCrimmon et al. (2015) explored the safety and feasibility of a foot-drop-targeted BCI-FES physiotherapy in chronic stroke survivors. The authors concluded that BCI-FES therapy was safe and that a large number of subjects experienced improvements in terms of gait speed and dorsiflexion aROM.

Optimal treatment protocols for the lower limb and gait recovery are difficult to elicit due to the inconsistent and wide ranging outcome measures, varying modalities to NMES, and the different stimulation parameters used (Kafri et al., 2015). Kafri et al. (2015) reported that in the case of reviewed studies, frequency values generally ranged between 15-50 Hz, whereas pulse duration varied between 280-400 μs in line with a current amplitude value of 4-85 mA. The authors indicated that the stimulation parameters were not reported in a consistent manner, stating that while some studies did not report any parameters, others varied in the amount of detail provided. Kesar et al. (2010) compared knee and ankle kinematics during the swing phase of gait when FES was delivered to the ankle dorsiflexors using variable frequency trains (VFTs) versus constant-frequency trains (CFTs). A 30 Hz constant-frequency train was applied in the case of FES using CFTs, whereas a 3-pulse burst of high frequency (200 Hz) followed by a lower frequency (30 Hz) was applied in the case of FES using VFTs. The authors concluded that novel FES systems capable of delivering VFTs during gait can result in an enhanced drop-foot correction as opposed to the traditional FES systems that deliver CFTs. The results also suggested that the delivery timing of FES during gait was critical and that it is also required further investigation. Concerning doses, Smith et al. (2003) have demonstrated a dose-response relationship between FES and brain activation in sensory and motor regions contralateral to the stimulation and it is also suggested that more intensive training may be beneficial (Bosch et al., 2014). In relation to the electrical stimulation orthotic substitute devices, Bosch et al. (2014) pointed out that the limited total intervention time, small sample size, and lack of focus on gait training raise questions about the optimal interventions and training intensity needed to obtain beneficial effects.

There is evidence supporting the use of electrical stimulation for therapeutic training and as a replacement or alternative to an AFO (Bosch et al., 2014). These positive effects may include changes in gait characteristics and body function variables (Kafri et al., 2015). The therapeutic effect of FES on balance did not demonstrate any clear patterns. In addition, consistent findings indicate that when FES is used as an alternative to an assistive device it has no superior therapeutic effects than AFO (Kafri et al., 2015; Bosch et al., 2014).

In terms of patient characteristics, it is not clear which individuals will benefit most from NMES, and what baseline characteristics predict better therapeutic outcomes (Kafri et al., 2015). Merletti et al. (1979) suggested that the time since lesion, spasticity, and intensity of treatment are important predictors of positive therapeutic results.

The use of FES combined with different walking strategies has been shown to result in improvements in hemiplegic gait (Belda-Lois et al., 2011) These methods include cycling, leg movements in a side lying position, treadmill training, electromechanical gait training and so on. FES has also been applied in conjunction with mirror therapy indicating good results in gait ability, muscle strength and balance in hemiplegic stroke survivors (Ji et al., 2014; Lee et al., 2016). Preliminary studies have been carried out in relation to a multi-level NMES paradigm with a virtual reality-based adaptive response technology for post-stroke balance rehabilitation (Dutta et al., 2014). NMES can also be used along or as a part of a neuro-robot (Moreno et al., 2011), yet this evidence is still limited. Future research topics have been suggested in relation to the combination of FES and Lokomat-assisted robotic rehabilitation therapy that might lead to further mobility improvements (Jezernik et al., 2003).

Locomotion results from intricate dynamic interactions between a central programme and feedback mechanisms (Belda-Lois et al., 2011). While functional changes in mobility performance may be attributed to the direct effects on muscle structure and function, two other central mechanisms may be involved namely the direct effects of electrical stimulation on brain plasticity and the enhanced motor learning resulting from repetitive practice (Kafri et al., 2015). Findings indicate that NMES in conjunction with voluntary muscle contractions is superior to individual NMES in terms of cortical excitability and plasticity (Kafri et al., 2015).

HEMIPLEGIC SHOULDER

The hemiplegic shoulder constitutes a challenge in the rehabilitation of stroke patients. Shoulder subluxation may affect the upper limb treatment process. It can also lead to additional complications, such as pain and delays in functional recovery (Vafadar et al., 2015; Gu et al., 2016).

Various meta-analyses and reviews have been published on this subject in recent years. Sheffler et al. (2007) argued that the use of NMES appears to reduce shoulder subluxation along with the pain-free ROM, as well as facilitate motor recovery, especially in the case of acute stroke patients. However, additional studies are required in order to clarify the clinical efficacy thereof. The conclusions formulated by Vafadar et al. (2015) and Gu et al. (2016) indicated similar findings. Both authors pointed out that NMES could be used in order to prevent or reduce shoulder subluxation early after a stroke, yet findings did not support their efficacy for pain reduction, improvement in arm strength, movement, functional activity, and quality of life following a stroke.

The NMES protocols applied to the hemiplegic shoulder are very disparate and clinical trials do not fully describe them in the majority of cases, thus making it difficult for a comparison to be made between the studies. The stimulated musculature, which includes the supraspinatus and the posterior deltoid, is the only variable that most studies have in common, whereas the rest of the stimulation parameters vary from one author to another. The frequency ranges mainly between 10 and 30 Hz, the most frequent pulse duration is 300 μs, although on many occasions these values are not defined by the author. The waveform as along with the intensity parameters also fail to be indicated in the vast majority of studies. Furthermore, ramping up/down periods and duty cycles are very different in the defined protocols, which is also the case for the treatment doses.

More extensive and accurate studies should be carried out in order to reach conclusive results about the long-term effects of NMES and their relation to the post-stroke time period (Vafadar et al., 2015; Gu et al., 2016).

DYSPHAGIA

Dysphagia or swallowing problems are common and they occur in 45-65% of patients following an acute stroke (Daniels et al., 1998; Lee et al., 2014). Complications of dysphagia include aspiration leading to chest infections and pneumonia, malnutrition, inability to rehabilitate, prolonged length of hospitalisation, and an increased risk of death (Geeganage et al., 2012). Different interventions are applied in order to accelerate the recovery of the swallowing function. NMES for the treatment of dysphagia is a relatively new therapeutic method (Poorjavad et al., 2014; Tan et al., 2013; Park et al., 2012). This new approach has generated a lot of interest across a vast amount of publications in recent years, yet the available evidence is limited (Chen et al., 2016; Poorjavad et al., 2014).

A recent meta-analysis (Tan et al., 2013) evaluated the effect of NMES versus traditional therapy (TT) in patients with dysphagia of different aetiology. In the subgroup analysis of stroke patients, the effectiveness of NMES and TT was comparable, patients with dysphagia may benefit to a greater extent from NMES when combined with TT. Following the same approach, the latest meta-analysis (Chen et al., 2016) concluded that swallowing treatment by means of NMES seems to be more effective than the interventions without it in the case of post-stroke dysphagia in the short term, and the authors remarked that the evidence is insufficient in order to indicate whether NMES alone is superior to swallowing therapy.

Related to the NMES protocol, after reviewing various clinical trials, the device predominantly used is the Vitalstim® equipment (Chattanooga Group, Hixson, TN, USA, device approved by the FDA for the treatment of dysphagia), including biphasic pulses, a frequency of 80 Hz and a pulse duration of 700 μs. The stimulation intensity is adjusted according to the sensory or motor threshold, the usual dosage of treatment is 60 min per day for 5 days per week, with the duration of the study fluctuating between 5 days and 6 weeks. On the same subject, a review (Poorjavad et al., 2014) related to the dysphagic patients following a stroke, analysed the intensity of stimulation (sensory or motor level) and the electrode placement at the level of the neck in more detail. The authors pointed out that these two parameters of the protocol could have important effects on the outcomes and they had not been extensively considered by previously published reviews (Steele et al., 2007; Clark et al., 2009; Tan et al., 2013). Generally, NMES on the swallowing muscles has been applied with two general purposes, namely to cause muscle contractions in which case the intensity of the electrical current is increased until the muscle contraction occurs, and to stimulate the sensory pathways in which case the sensory threshold is usually identified as the lowest current level at which the patient feels a tingling sensation in the neck skin. Electrode placement usually involves the suprahyoid or infrahyoid muscles. According to this review, the majority of the studies reported some positive treatment outcomes in the case of the electrical stimulation to the neck musculature in relation to the swallowing function of stroke patients, especially when the intensity of the stimulus was adjusted at the sensory level or when the motor electrical stimulation was applied to the infrahyoid muscles when swallowing. Nevertheless, additional studies ought to be carried out in relation to the physiologic and neurologic effects of these therapeutic methods, which would help to design specific and individual treatment protocols. Rofes et al. (2013) compared the efficacy and safety of 10 days of treatment with NMES being recorded at sensory and motor intensities in chronic post-stroke patients. The authors concluded that NMES improved the swallow response and safety of swallow both at sensory and motor intensities, and the efficacy of swallow following the motor treatment. A more significant effect was observed when the stimulus was applied at motor intensities.

The mechanisms underlying the efficacy of electrical stimulation in the treatment of post-stroke dysphagia are not well-known. Gallas et al. (2010) reported that the plasticity of the sensory swallowing cortex is presumed to be based on the long-term reorganisation of the human cortex driven by the sensory stimulation. Rofes et al. (2013) and Chen et al. (2016) suggested that both peripheral and central modifications could be responsible for the effect of the therapy, whereas Sun et al. (2013) mentioned that NMES could improve the hyolaryngeal elevation, restore motor function of weak muscles, combat disuse atrophy, enhance sensory awareness and facilitate muscle contractions.

URINARY INCONTINENCE

Urinary incontinence (UI) following a stroke affects more than one-third of acute stroke patients (Liu et al., 2016), namely between 40% and 60% of patients who are hospitalised, 25% of patients upon hospital discharge, and 15% of patients one year after stroke (Thomas et al., 2011). NMES has also been proposed as a treatment tool for the patients with post-stroke urinary symptoms (Skeil et al., 2001; Cooperberg et al., 2005), although there is a paucity of existing research about it. The rehabilitation of patients with post-stroke UI is still a relatively new topic, and its occurrence has not been identified as a problem by therapists, even in developed countries (Guo et al., 2014).

Guo et al. (2014) investigated the therapeutic effect of TENS on post-stroke UI. The treatment group received therapy by means of TENS for 30 minutes once a day over the course of 60 days, at a pulse duration of 70 μs and a frequency of 75 Hz. In terms of the electrode placement, the positive electrode was placed on the second lumbar spinous process, whereas the two negative electrodes were located on the inside of the middle and the lower third of the junction between the posterior superior iliac spine and the ischia node. The daily micturition, nocturia, urgent urination, and urge UI in the treatment group improved significantly compared to the control group. Later, the same authors published another related article (Liu et al., 2016), comparing two protocols in which TENS was applied at different frequencies, namely 20 Hz and 75 Hz, and pulse durations of 150 μs, for 30 min once per day over the course of 90 days. This time, two positive electrodes were used, which were placed in the region of the second sacral level on opposite sides of the vertebral column, whereas the negative electrodes were placed in the same manner as the previously published article. The stroke patients treated with a frequency of 20 Hz had superior Overactive Bladder Symptom Scores, Barthel Index totals, urodynamic values and voiding diary parameters. Despite obtaining promising results, the authors recommend that more studies involving larger sample sizes and longer follow-up periods ought to be conducted in order to confirm the long-term effects of electrical stimulation.

Visuo-Spatial Neglect

Visuo-spatial neglect is a neurological disorder that often follows a stroke, especially after sustaining damage to the right hemisphere of the brain. It refers to the defective ability of patients to explore the side of the space contralateral to the brain lesion and to report stimuli presented in that portion of the space (Vallar et al., 1998). Different types of interventions are suggested. Electrical stimulation is investigated in some studies, although the use thereof is not very widespread and well-known.

There are few clinical trials in the case of this specific population. Many of these investigations include samples with other neurological disorders that also sustain brain lesions. Vallar et al. (1995) reported that TENS applied to the muscles on the left side of the neck led to a temporary improvement in 93% of neglect patients due to different neurological disorders. Left-hand stimulation had a comparable effect. Schröder et al. (2008) recommend TENS as a possible visual neglect therapy based on the results of their research. They compared different treatments, namely exploration training alone and in combination with TENS or optokinetic stimulation (OKS). In the case of TENS group, the stimulation frequency was set at 100 Hz with a pulse duration of 100 μs (as in the study of Vallar et al., 1995). The positive electrode was placed over the left trapezius muscle halfway between the head and the shoulder and the negative one was attached about 40 mm below, on the upper left shoulder. The intensity was individually determined by increasing it gradually until a pleasant tingling in the neck was reported. Each patient received 20 therapy sessions each lasting between 25 and 40 min over a period of 4 weeks. The results of this RCT indicated that the groups receiving TENS or OKS showed significant improvements in their reading and writing abilities, although the improvement in neglect test scores disappeared during the follow-up period in TENS group. Pérennou et al. (2001) demonstrated the manner in which it is possible to modulate the postural instability in neglect patients by means of an appropriate somatosensory manipulation, namely the use of TENS. The authors concluded that TENS stimulation applied at the neck level could be a convenient tool for inducing postural neuromodulation in a clinical context. However, the duration of the TENS efficiency with reference to the postural stability lasted for a short period of time. In this regard, further investigation is needed to determine whether repeated applications of TENS eventually associated with some active retraining techniques will facilitate the functional restoration of neglect patients after a stroke.

The specific mechanisms underlying this intervention are complex and unclear, but the latter findings suggest that TENS may elicit a proprioceptive input through large diameter afferents (Vallar et al., 1995). Vallar et al. (1995) suggested that the improvement could be attributed to a resulted from a non-specific activation of the right hemisphere or a directional effect on the egocentric coordinates of the extrapersonal space. The activation of the neural systems involved in the organization of the egocentric representation has also been suggested by Schröder et al. (2008) and Pizzamiglio et al. (1996). Lomber and Payne (1996) regarded the unilateral neglect as the result of an imbalance in the activity of the two hemispheres at the cortical or subcortical level. In this manner, TENS might serve to reduce this imbalance by increasing the activity of the right hemisphere.

CONCLUSION

Advances in neuroscience and the development of new technologies will possibly give rise to new applications of NMES for upper limb recovery after a stroke. Future NMES protocols should take into account various aspects including bilateral activity, repetitive movement, maximum subject attentiveness, and tasks that motivate goal-driven motor skills. These variables are suspected to promote motor relearning and to enhance the recovery of upper limb in stroke patients. NMES could be a complement to rehabilitation programmes along with other emerging therapies. Further research is needed to establish whether the parameters, treatment doses and patient characteristics are crucial for the effectiveness thereof. Such research should also elucidate the specific mechanism of action associated to the different stimulation modalities in order to select the most suitable patients and to optimise individual treatment protocols to achieve larger treatment effects.

Further research is also required in order to investigate the neurobiological principles that underlie any functional changes in lower limb and gait, to elicit the subpopulation of stroke patients that will most benefit therefrom and to evaluate their effectiveness in line with other emerging strategies. These studies would also be useful in order to optimise the protocols applied. Lower limb and gait programmes should include training principles such as goal-directed, task-specific training, intensive in terms of the number of repetitions, challenging, and motivating, in an attempt to facilitate motor learning on the basis of neural plasticity. In addition, further technological advances are expected to increase the availability and usability of gait FES systems, while in turn significantly contributing to gait and quality of life of post-stroke patients. These new systems should be associated with cost effectiveness studies in order to assess its viability in clinical practice.

In relation to the hemiplegic shoulder, It would be interesting to carry out RCTs with the aim of comparing different treatment protocols in order to determine which parameters are the most effective, as well as to consider whether the treatment programme could be different depending on the patient characteristics. An accurate assessment of the shoulder to be treated and a more homogeneous selection of the sample could provide more information about the effectiveness of NMES on the hemiplegic shoulder and establish protocols in accordance with the treatment purpose.

With reference to the future lines of research in relation to post-stroke dysphagia, there is a need to carry out future controlled studies with larger group of patients and a long-term follow-up period. New findings could thus be used in order to optimise the electrical stimulation parameters, to investigate the role of lesion sites and time from stroke, and to study the neurophysiologic measurements of cortical excitability with the purpose of confirming the action mechanism of the therapy.

Future research is needed to obtain more concluding results in the case of novel strategies associated to urinary incontinence and visuo-spatial neglect in order to establish effective treatments.

It is therefore important to possess extensive knowledge about the effects of NMES in these different areas and to further investigate their maintenance during the follow-up period, in order to guide clinical decisions and the bioengineering community in the development of appropriate technologies. New advances in nanotechnology, e-textiles and telerehabilitation will play an interesting role in future stroke rehabilitation. A combination of different rehabilitation strategies seems to be more effective as opposed to the individual implementation thereof. Future research would make it possible to analyse the impact of rehabilitation on brain plasticity, in order to adapt treatment resources to meet the needs of each patient and to subsequently optimise the recovery process. More rigorous clinical and economic evidence is needed to incentivise practitioners and healthcare decision-makers to implement these NMES treatments. In addition, a multidisciplinary interactive effort between engineering, neuroscience and medicine is required to progress in the field of neurorehabilitation.

REFERENCES

Alon G, Levitt AF, McCarthy PA. Functional electrical stimulation may modify the poor prognosis of stroke survivors with severe motor loss of the upper extremity: a preliminary study. *Am J Phys Med Rehabil*. 2008; 87: 627-36.

Belda-Lois JM, Mena-del Horno S, Bermejo-Bosch I, Moreno JC, Pons JL, Farina D, et al. Rehabilitation of gait after stroke: a review towards a top-down approach. *J Neuroeng Rehabil*. 2011; 8: 66.

Bermejo F, Iglesias P, Calleja P, Ruiz PM. [Opinions on the organisation of health care in strokes. A survey among hospital medical directors in Spain]. *Rev Neurol*. 2009; 48: 395-9.

Biasiucci A, Leeb R, Al-Khodairy A, Buhlmann A, Millán R. Motor recovery after stroke by means of BCI-guided functional electrical stimulation. Proceedings of the Fifth International Brain-Computer Interface Meeting 2013. 2013 June 3-7; Asilomar, California, US.

Bosch PR, Harris JE, Wing K. American Congress of Rehabilitation Medicine (ACRM) Stroke Movement Interventions Subcommittee. Review of therapeutic electrical stimulation for dorsiflexion assist and orthotic substitution from the American Congress of Rehabilitation Medicine stroke movement interventions subcommittee. *Arch Phys Med Rehabil*. 2014; 95: 390-6.

Boyaci A, Topuz O, Alkan H, Ozgen M, Sarsan A, Yildiz N, et al. Comparison of the effectiveness of active and passive neuromuscular electrical stimulation of hemiplegic upper extremities: a randomized, controlled trial. *Int J Rehabil Res*. 2013; 36: 315-22.

Cauraugh JH, Kim S. Two coupled motor recovery protocols are better than one electromyogram-triggered neuromuscular stimulation and bilateral movements. *Stroke*. 2002; 33: 1589-94.

Cauraugh JH, Kim SB, Duley A. Coupled bilateral movements and active neuromuscular stimulation: intralimb transfer evidence during bimanual aiming. *Neurosci Lett*. 2005; 382: 39-44.

Cauraugh JH, Kim SB. Chronic stroke motor recovery: duration of active neuromuscular stimulation. *J Neurol Sci*. 2003; 215: 13-9.

Chae J, Sheffler L, Knutson J. Neuromuscular electrical stimulation for motor restoration in hemiplegia. *Top Stroke Rehabil.* 2008; 15: 412-26.

Chen YW, Chang KH, Chen HC, Liang WM, Wang YH, Lin YN. The effects of surface neuromuscular electrical stimulation on post-stroke dysphagia: a systemic review and meta-analysis. *Clin Rehabil.* 2016; 30: 24-35.

Chung E, Park SI, Jang YY, Lee BH. Effects of brain-computer interface-based functional electrical stimulation on balance and gait function in patients with stroke: preliminary results. *J Phys Ther Sci.* 2015; 27: 513-6.

Clair-Auger JM, Collins DF, Dewald JPA. The effects of wide pulse neuromuscular electrical stimulation on elbow flexion torque in individuals with chronic hemiparetic stroke. *Clin Neurophysiol.* 2012; 123: 2247-55.

Clark H, Lazarus C, Arvedson J, Schooling T, Frymark T. Evidence-based systematic review: effects of neuromuscular electrical stimulation on swallowing and neural activation. *Am J Speech Lang Pathol.* 2009; 18: 361-75.

Conforto AB, Ferreiro KN, Tomasi C, dos Santos RL, Moreira VL, Marie SK, et al. Effects of somatosensory stimulation on motor function after subacute stroke. *Neurorehabil Neural Repair.* 2010; 24: 263-72.

Cooperberg MR, Stoller ML. Percutaneous neuromodulation. *Urol Clin North Am.* 2005; 32: 71-8.

Daly JJ, Cheng R, Rogers J, Litinas K, Hrovat K, Dohring M. Feasibility of a new application of noninvasive Brain Computer Interface (BCI): a case study of training for recovery of volitional motor control after stroke. *J Neurol Phys Ther.* 2009; 33: 203-11.

Daly JJ, Hogan N, Perepezko EM, Krebs HI, Rogers JM, Goyal KS, et al. Response to upper-limb robotics and functional neuromuscular stimulation following stroke. *J Rehabil Res Dev.* 2005; 42: 723-36.

Daniels SK, Brailey K, Priestly DH, Herrington LR, Weisberg LA, Foundas AL. Aspiration in patients with acute stroke. *Arch Phys Med Rehabil.* 1998; 79: 14-9.

de Kroon JR, Ijzerman MJ, Chae J, Lankhorst G, Zilvold G. Relation between stimulation characteristics and clinical outcome in studies using electrical stimulation to improve motor control of the upper extremity in stroke. *J Rehabil Med.* 2005; 37: 65-74.

de Kroon JR, IJzerman MJ, Lankhorst GJ, Zilvold G. Electrical stimulation of the upper limb in stroke: stimulation of the extensors of the hand vs. alternate stimulation of flexors and extensors. *Am J Phys Med Rehabil.* 2004; 83: 592-600.

de Kroon JR, IJzerman MJ. Electrical stimulation of the upper extremity in stroke: cyclic versus EMG-triggered stimulation. *Clin Rehabil.* 2008; 22: 690-7.

Díaz L, Arellano ME, del Valle MG, Miranda A, Rodríguez G, Montero A. [Usefulness of neuromuscular electrical stimulation after the application of botulinum toxin in children with spastic hemiparesis]. *Rev Mex Med Fis Rehab.* 2005; 17: 16-22.

Doucet BM, Griffin L. High-versus low-frequency stimulation effects on fine motor control in chronic hemiplegia: a pilot study. *Top Stroke Rehabil.* 2013; 20: 299-307.

Dutta A, Lahiri U, Das A, Nitsche MA, Guiraud D. Post-stroke balance rehabilitation under multi-level electrotherapy: a conceptual review. *Front Neurosci.* 2014; 8: 403.

Eng JJ, Tang PF. Gait training strategies to optimize walking ability in people with stroke: a synthesis of the evidence. *Expert Rev Neurother.* 2007; 7: 1417-36.

Evers SM, Struijs JN, Ament AJ, van Genugten ML, Jager JC, van den Bos GA. International comparison of stroke cost studies. *Stroke.* 2004, 35: 1209-15.

Farmer SE, Durairaj V, Swain I, Pandyan AD. Assistive technologies: can they contribute to rehabilitation of the upper limb after stroke? *Arch Phys Med Rehabil.* 2014; 95: 968-85.

Gallas S, Marie JP, Leroi AM, Verin E. Sensory transcutaneous electrical stimulation improves post-stroke dysphagic patients. *Dysphagia.* 2010; 25: 291-7.

Geeganage C, Beavan J, Ellender S, Bath PM. Interventions for dysphagia and nutritional support in acute and subacute stroke. *Cochrane Database Syst Rev.* 2012; 10: CD000323.

Glanz M, Klawansky S, Stason W, Berkey C, Chalmers TC. Functional electrostimulation in poststroke rehabilitation: a meta-analysis of the randomized controlled trials. *Arch Phys Med Rehabil.* 1996; 77: 549-53.

Gu P, Ran JJ. Electrical Stimulation for Hemiplegic Shoulder Function: a systematic review and meta-analysis of 15 randomized controlled trials. *Arch Phys Med Rehabil.* 2016; 97: 1588-94.

Guo ZF, Liu Y, Hu GH, Liu H, Xu YF. Transcutaneous electrical nerve stimulation in the treatment of patients with poststroke urinary incontinence. *Clin Interv Aging.* 2014; 9: 851-6.

Hara Y, Ogawa S, Tsujiuchi K, Muraoka Y. A home-based rehabilitation program for the hemiplegic upper extremity by power-assisted functional electrical stimulation. *Disabil Rehabil.* 2007; 11: 1-9.

Hara Y. Neurorehabilitation with new functional electrical stimulation for hemiparetic upper extremity in stroke patients. *J Nippon Med Sch*. 2008; 75: 4-14.

Hemmen B, Seelen HA. Effects of movement imagery and electromyography-triggered feedback on arm hand function in stroke patients in the subacute phase. *Clin Rehabil*. 2007; 21: 587-94.

Hong IK, Choi JB, Lee JH. Cortical changes after mental imagery training combined with electromyography-triggered electrical stimulation in patients with chronic stroke. *Stroke*. 2012; 43: 2506-9.

Howlett OA, Lannin NA, Ada L, McKinstry C. Functional electrical stimulation improves activity after stroke: a systematic review with meta-analysis. *Arch Phys Med Rehabil*. 2015; 96: 934-43.

Hsu SS, Hu MH, Wang YH, Yip PK, Chiu JW, Hsieh CL. Dose-response relation between neuromuscular electrical stimulation and upper-extremity function in patients with stroke. *Stroke*. 2010; 41: 821-4.

Hu XL, Tong RK, Ho NS, Xue JJ, Rong W, Li LS. Wrist rehabilitation assisted by an electromyography-driven neuromuscular electrical stimulation robot after stroke. *Neurorehabil Neural Repair*. 2015; 29: 767-76.

IJzerman MJ, Renzenbrink GJ, Geurts AC. Neuromuscular stimulation after stroke: from technology to clinical deployment. *Expert Rev Neurother*. 2009; 9: 541-52.

Jezernik S, Colombo G, Keller T, Frueh H, Morari M. Robotic orthosis lokomat: a rehabilitation and research tool. *Neuromodulation*. 2003; 6: 108-15.

Ji SG, Cha HG, Kim MK, Lee CR. The effect of mirror therapy integrating functional electrical stimulation on the gait of stroke patients. *J Phys Ther Sci*. 2014; 26: 497-9.

Kafri M, Laufer Y. Therapeutic effects of functional electrical stimulation on gait in individuals post-stroke. *Ann Biomed Eng*. 2015; 43: 451-66.

Kesar TM, Perumal R, Jancosko A, Reisman DS, Rudolph KS, Higginson JS, et al. Novel patterns of functional electrical stimulation have an immediate effect on dorsiflexor muscle function during gait for people poststroke. *Phys Ther*. 2010; 90: 55-66.

Kim JH, Lee BH. Mirror therapy combined with biofeedback functional electrical stimulation for motor recovery of upper extremities after stroke: a pilot randomized controlled trial. *Occup Ther Int*. 2015; 22: 51-60.

Kim T, Kim S, Lee B. Effects of action observational training plus brain-computer interface-based functional electrical stimulation on paretic arm motor recovery in patient with stroke: a randomized controlled trial. *Occup Ther Int*. 2016; 23: 39-47.

Kimberley TJ, Lewis SM, Auerbach EJ, Dorsey LL, Lojovich JM, Carey JR. Electrical stimulation driving functional improvements and cortical changes in subjects with stroke. *Exp Brain Res.* 2004; 154: 450-60.

Knutson JS, Fu MJ, Sheffler LR, Chae J. Neuromuscular electrical stimulation for motor restoration in hemiplegia. *Phys Med Rehabil Clin N Am.* 2015; 26: 729-45.

Knutson JS, Gunzler DD, Wilson RD, Chae J. Contralaterally controlled functional electrical stimulation improves hand dexterity in chronic hemiparesis: a randomized trial. *Stroke.* 2016; 47: 2596-602.

Knutson JS, Harley MY, Hisel TZ, Chae J. Improving hand function in stroke survivors: a pilot study of contralaterally controlled functional electric stimulation in chronic hemiplegia. *Arch Phys Med Rehabil.* 2007; 88: 513-20.

Knutson JS, Harley MY, Hisel TZ, Makowski NS, Fu MJ, Chae J. Contralaterally controlled functional electrical stimulation for stroke rehabilitation. *Conf Proc IEEE Eng Med Biol Soc.* 2012; 2012: 314-7.

Kowalczewski J, Gritsenko V, Ashworth N, Ellaway P, Prochazka A. Upper-extremity functional electric stimulation-assisted exercises on a workstation in the subacute phase of stroke recovery. *Arch Phys Med Rehabil.* 2007; 88: 833-9.

Lee D, Lee G, Jeong J. Mirror therapy with neuromuscular electrical stimulation for improving motor function of stroke survivors: a pilot randomized clinical study. *Technol Health Care.* 2016; 24: 503-11.

Lee KW, Kim SB, Lee JH, Lee SJ, Ri JW, Park JG. The effect of early neuromuscular electrical stimulation therapy in acute/subacute ischemic stroke patients with dysphagia. *Ann Rehabil Med.* 2014; 38: 153-9.

Liberson WT, Holmquest HJ, Scot D, Dow M. Functional electrotherapy: stimulation of the peroneal nerve synchronized with the swing phase of the gait of hemiplegic patients. *Arch Phys Med Rehabil.* 1961; 42: 101-5.

Lin KC, Huang PC, Chen YT, Wu CY, Huang WL. Combining afferent stimulation and mirror therapy for rehabilitating motor function, motor control, ambulation, and daily functions after stroke. *Neurorehabil Neural Repair.* 2014; 28: 153-62.

Lindquist AR, Prado CL, Barros RM, Mattioli R, da Costa PH, Salvini TF. Gait training combining partial body-weight support, a treadmill, and functional electrical stimulation: effects on poststroke gait. *Phys Ther.* 2007, 87: 1144-54.

Liu Y, Xu G, Luo M, Teng HF. Effects of transcutaneous electrical nerve stimulation at two frequencies on urinary incontinence in poststroke patients: a randomized controlled trial. *Am J Phys Med Rehabil.* 2016; 95: 183-93.

Lomber SG, Payne BR. Removal of two halves restores the whole: reversal of visual hemineglect during bilateral cortical or collicular inactivation in the cat. *Visual Neuroscience.* 1996; 13: 1143-56.

Lourenção MI, Battistella LR, Martins LC, Litvoc J. Analysis of the results of functional electrical stimulation on hemiplegic patients' upper extremities using the Minnesota manual dexterity test. *Int J Rehabil Res.* 2005; 28: 25-31.

Mangold S, Schuster C, Keller T, Zimmermann-Schlatter A, Ettlin T. Motor training of upper extremity with functional electrical stimulation in early stroke rehabilitation. *Neurorehabil Neural Repair.* 2009; 23: 184-90.

McCabe J, Monkiewicz M, Holcomb J, Pundik S, Daly JJ. Comparison of robotics, functional electrical stimulation, and motor learning methods for treatment of persistent upper extremity dysfunction after stroke: a randomized controlled trial. *Arch Phys Med Rehabil.* 2015; 96: 981-90.

McCrimmon CM, King CE, Wang PT, Cramer SC, Nenadic Z, Do AH. Brain-controlled functional electrical stimulation therapy for gait rehabilitation after stroke: a safety study. *J Neuroeng Rehabil.* 2015; 12: 57.

Meadmore KL, Exell TA, Hallewell E, Hughes AM, Freeman CT, Kutlu M, et al. The application of precisely controlled functional electrical stimulation to the shoulder, elbow and wrist for upper limb stroke rehabilitation: a feasibility study. *J Neuroeng Rehabil.* 2014; 11: 105.

Mehrholz J, Elsner B, Werner C, Kugler J, Pohl M. Electromechanical-assisted training for walking after stroke: updated evidence. *Stroke.* 2013; 44: e127-8.

Merletti R, Andina A, Galante M, Furlan I. Clinical experience of electronic peroneal stimulators in 50 hemiparetic patients. *Scand J Rehabil Med.* 1979; 11: 111-21.

Millán JD, Rupp R, Müller-Putz GR, Murray-Smith R, Giugliemma C, Tangermann M, et al. Combining brain-computer interfaces and assistive technologies: state-of-the-art and challenges. *Front Neurosci.* 2010; 4: pii 161.

Moreno JC, Del Ama AJ, de Los Reyes-Guzmán A, Gil-Agudo A, Ceres R, Pons JL. Neurorobotic and hybrid management of lower limb motor disorders: a review. *Med Biol Eng Comput.* 2011; 49: 1119-30.

Müller-Putz GR, Scherer R, Pfurtscheller G, Rupp R. EEG-based neuroprosthesis control: a step towards clinical practice. *Neurosci Lett.* 2005; 382: 169-74.

Nakayama H, Jørgensen HS, Raaschou HO, Olsen TS. Recovery of upper extremity function in stroke patients: the Copenhagen Stroke Study. *Arch Phys Med Rehabil.* 1994; 75: 394-8.

Nascimento LR, Michaelsen SM, Ada L, Polese JC, Teixeira-Salmela LF. Cyclical electrical stimulation increases strength and improves activity after stroke: a systematic review. *J Physiother.* 2014; 60: 22-30.

O'Connor RJ, Jackson A, Makower SG, Cozens A, Levesley M. A proof of concept study investigating the feasibility of combining iPAM robot assisted rehabilitation with functional electrical stimulation to deliver whole arm exercise in stroke survivors. *J Med Eng Technol.* 2015; 39: 411-8.

Page SJ, Levin L, Hermann V, Dunning K, Levine P. Longer versus shorter daily durations of electrical stimulation during task-specific practice in moderately impaired stroke. *Arch Phys Med Rehabil.* 2012; 93: 200-6.

Page SJ, Levine P. Back from the brink: electromyography-triggered stimulation combined with modified constraint-induced movement therapy in chronic stroke. *Arch Phys Med Rehabil.* 2006; 87: 27-31.

Pérennou DA, Leblond C, Amblard B, Micallef JP, Hérisson C, Pélissier JY. Transcutaneous electric nerve stimulation reduces neglect-related postural instability after stroke. *Arch Phys Med Rehabil.* 2001; 82: 440-8.

Pizzamiglio L, Vallar G, Magnotti L. Transcutaneous electrical stimulation of the neck muscles and hemineglect rehabilitation. *Restor Neurol Neurosci.* 1996; 10: 197-203.

Pomeroy VM, King L, Pollock A, Baily-Hallam A, Langhorne P. Electrostimulation for promoting recovery of movement or functional ability after stroke. *Cochrane Database Syst Rev.* 2006; 2: CD003241.

Poorjavad M, Talebian Moghadam S, Nakhostin Ansari N, Daemi M. Surface electrical stimulation for treating swallowing disorders after stroke: a review of the stimulation intensity levels and the electrode placements. *Stroke Res Treat.* 2014; 2014: 918057.

Popovic MB, Popovic DB, Sinkjær T, Stefanovic A, Schwirtlich L. Clinical evaluation of functional electrical therapy in acute hemiplegic subjects. *J Rehabil Res Dev.* 2003; 40: 443-53.

Popovic MB, Popovic DB, Sinkjaer T, Stefanovic A, Schwirtlich L. Restitution of reaching and grasping promoted by functional electrical therapy. *Artif Organs.* 2002; 26: 271-5.

Powell J, Pandyan AD, Granat M, Cameron M, Stott DJ. Electrical stimulation of wrist extensors in poststroke hemiplegia. *Stroke*. 1999; 30: 1384-9.

Quandt F, Hummel FC. The influence of functional electrical stimulation on hand motor recovery in stroke patients: a review. *Exp Transl Stroke Med*. 2014; 6: 9.

Rigau D, Álvarez J, Gil A, Abilleira S, Borras FX, Armario P, et al. [Primary and secondary prevention of stroke: a guideline]. *Med Clin (Barc)*. 2009; 133: 754-62.

Ring H, Rosenthal N. Controlled study of neuroprosthetic functional electrical stimulation in sub-acute post-stroke rehabilitation. *J Rehabil Med*. 2005; 37: 32-6.

Rodríguez JM. [Electrotherapy in physiotherapy]. 2ª ed. Madrid: Médica Panamericana; 2004.

Rofes L, Arreola V, López I, Martin A, Sebastián M, Ciurana A, et al. Effect of surface sensory and motor electrical stimulation on chronic poststroke oropharyngeal dysfunction. *Neurogastroenterol Motil*. 2013; 25: 888-e701.

Rosewilliam S, Malhotra S, Roffe C, Jones P, Pandyan AD. Can surface neuromuscular electrical stimulation of the wrist and hand combined with routine therapy facilitate recovery of arm function in patients with stroke? *Arch Phys Med Rehabil*. 2012; 93: 1715-21.

Schröder A, Wist ER, Hömberg V. TENS and optokinetic stimulation in neglect therapy after cerebrovascular accident: a randomized controlled study. *Eur J Neurol*. 2008; 15: 922-7.

Scutt P, Lee HS, Hamdy S, Bath PM. Pharyngeal electrical stimulation for treatment of poststroke dysphagia: individual patient data meta-analysis of randomised controlled trials. *Stroke Research and Treatment*. 2015; 2015: 429053.

Sentandreu Mañó T, Salom Terrádez JR, Tomás JM, Meléndez Moral JC, de la Fuente Fernández T, Company José C. [Electrical stimulation in the treatment of the spastic hemiplegic hand after stroke: a randomized study]. *Med Clin (Barc)*. 2011; 137: 297-301.

Sheffler LR, Chae J. Neuromuscular electrical stimulation in neurorehabilitation. *Muscle Nerve*. 2007; 35: 562-90.

Silvoni S, Ramos-Murguialday A, Cavinato M, Volpato C, Cisotto G, Turolla A, et al. Brain-computer interface in stroke: a review of progress. *Clin EEG Neurosci*. 2011; 42: 245-52.

Singer BJ, Vallence AM, Cleary S, Cooper I, Loftus AM. The effect of EMG triggered electrical stimulation plus task practice on arm function in chronic stroke patients with moderate-severe arm deficits. *Restor Neurol Neurosci.* 2013; 31: 681-91.

Smith GV, Alon G, Roys SR, Gullapalli RP. Functional MRI determination of a dose-response relationship to lower extremity neuromuscular electrical stimulation in healthy subjects. *Exp Brain Res.* 2003; 150: 33-9.

Soekadar SR, Birbaumer N, Slutzky MW, Cohen LG. Brain-machine interfaces in neurorehabilitation of stroke. *Neurobiol Dis.* 2015; 83: 172-9.

Steele CM, Thrasher AT, Popovic MR. Electric stimulation approaches to the restoration and rehabilitation of swallowing: a review. *Neurol Res.* 2007; 29: 9-15.

Sun SF, Hsu CW, Lin HS, Sun HP, Chang PH, Hsieh WL, et al. Combined neuromuscular electrical stimulation (NMES) with fiberoptic endoscopic evaluation of swallowing (FEES) and traditional swallowing rehabilitation in the treatment of stroke-related dysphagia. *Dysphagia.* 2013; 28: 557-66.

Tan C, Liu Y, Li W, Liu J, Chen L. Transcutaneous neuromuscular electrical stimulation can improve swallowing function in patients with dysphagia caused by non-stroke diseases: a meta-analysis. *J Oral Rehabil.* 2013; 40: 472-80.

Thomas LH, Watkins CL, French B, Sutton C, Forshaw D, Cheater F, et al. Study protocol: ICONS: identifying continence options after stroke: a randomised trial. *Trials.* 2011; 12: 131.

Vafadar AK, Côté JN, Archambault PS. Effectiveness of functional electrical stimulation in improving clinical outcomes in the upper arm following stroke: a systematic review and meta-analysis. *Biomed Res Int.* 2015; 2015: 729768.

Vallar G, Rusconi ML, Barozzi S, Bernardini B, Ovadia D, Papagno C, et al. Improvement of left visuo-spatial hemineglect by left-sided transcutaneous electrical stimulation. *Neuropsychologia.* 1995; 33: 73-82.

Vallar G. Spatial hemineglect in humans. *Trends Cogn Sci.* 1998; 2: 87-97.

Venkatakrishnan A, Francisco GE, Contreras-Vidal JL. Applications of brain-machine interface systems in stroke recovery and rehabilitation. *Curr Phys Med Rehabil Rep.* 2014; 2: 93-105.

Wilson RD, Page SJ, Delahanty M, Knutson JS, Gunzler DD, Sheffler LR, et al. Upper-limb recovery after stroke: a randomized controlled trial comparing EMG-triggered, cyclic, and sensory electrical stimulation. *Neurorehabil Neural Repair.* 2016; 30: 978-87.

Yun GJ, Chun MH, Park JY, Kim BR. The synergic effects of mirror therapy and neuromuscular electrical stimulation for hand function in stroke patients. *Ann Rehabil Med.* 2011; 35: 316-21.

BIOGRAPHICAL SKETCH

Trinidad Sentandreu-Maño

Department of Physiotherapy, Faculty of Physiotherapy,
University of Valencia, Gascó Oliag 5, 46010 Valencia, Spain.
E-mail: trinidad.sentandreu@uv.es

Education:

Degree in Physiotherapy. Faculty of Physiotherapy, University of Valencia, Spain

Research and Professional Experience:

Assistant professor. Department of Physiotherapy, University of Valencia.

Predoctoral training grant for researchers. Conselleria de Educación, Comunidad Valenciana, Spain. Foundation for the Promotion of Health and Biomedical Research (FISABIO), Valencia, Spain. Project entitled "Effects of neuromuscular electrical stimulation in the treatment of the elderly patient's hand with spastic hemiplegia after a stroke". File number ACIF/2012/017 (2012-2014).

Grant for research studies in technology and health practices based on scientific evidence. Conselleria de Sanidad, Comunidad Valenciana, Spain. Research Foundation of the Doctor Peset University Hospital, Valencia, Spain. Project entitled "Effect of electrical stimulation applied on the spastic hand in the elderly patients following a stroke". File number 004/2010 (2010).

Publications:

Sentandreu-Mañó T, Salom Terrádez JR, Tomás JM, Company José C. [Evidence relating to the use of distal neuromuscular electrical stimulation in the recovery of stroke patients: a systematic review]. *Fisioterapia*. 2016; 38: 99-112.

Sentandreu Mañó T, Salom Terrádez JR, Tomás JM, Meléndez Moral JC, de la Fuente Fernández T, Company José C. [Electrical stimulation in the treatment of the spastic hemiplegic hand after stroke: a randomized study]. *Med Clin (Barc).* 2011; 137: 297-301.

Espí López GV, Sentandreu Mañó T, Colorado Lluch MI, Dueñas Moscardó L. [Effects of an oculo-cervical exercise program in adults on cervical spine mobility]. *Fisioterapia.* 2011; 33: 41-49.

Cebrià Iranzo MA, Sentandreu Mañó T, Babiera Ricart MC, Igual Camacho C. [Effectiveness of therapeutic physical exercise in patients with intermittent claudication due to peripheral arterial disease: a review]. *Fisioterapia.* 2010; 32: 172-82.

Cebrià Iranzo MA, Sentandreu Mañó T, Espí Lopez GV, Amérigo Mora ER, Garcia Lucerga MC, Igual Camacho C. [Cardiocirculatory Physiotherapy CD-ROM]. Valencia (Spain): Publications Service of Valencia University; 2008. ISBN 978-84-691-9531-4. Legal Deposit V-4877-2008.

In: Evoked Potentials and Electrical Stimulation ISBN: 978-1-53611-075-3
Editor: Timothy Eisenhower

Chapter 4

The Effect of Oxidative Stress on Brain Electrical Activity and its Repercussions on Sensory Organization in Geriatric Rhesus Monkeys in Captivity

Alejandra Ibáñez-Contreras[1,2,3,4], Braulio Hernández-Godínez[1,2,3,5], Gamaliel Mendoza-Cuevas[2,3], Ulalume Hernández-Arciga[4] and Mina Königsberg[4]

[1]Posgrado en Ciencias Biológicas y de la Salud,
Universidad Autónoma Metropolitana, Unidad Iztapalapa, CDMX, México
[2]Applied Research in Experimental Biomedicine S.A. de C.V.
(APREXBIO), Laboratorio de Primatología, CDMX, México
[3]Biología Integral para Vertebrados (BIOINVERT®),
Unidad de Experimentación Animal, Estado de México, México
[4]Depto. de Ciencias de la Salud,
Laboratorio de Bioenergética y envejecimiento celular,
Universidad Autónoma Metropolitana, Unidad Iztapalapa, CDMX, México
[5]Centro Nacional de Investigación en Instrumentación
e Imagenología Médica (CI3M), Universidad Autónoma
Metropolitana-Unidad Iztapalapa, CDMX, México

Abstract

In the last decades, nonhuman primates (NHP) have been widely used as biomodels to study cerebral aging in relation to oxidative stress (OS) and the way in which organisms respond to it, since it is known that during aging reactive oxygen and nitrogen species (ROS and RNS) increase and, at the same time, the activity of the antioxidant systems decreases. In this sense, it has been described that when the neuronal plasma membrane components are oxidized, important neurophysiological changes occur, which potentially alter a great variety of electrophysiological parameters, generating a decrease in the nerve conduction velocity. During the last decades, it has been shown that with age increment, different mechanisms are generated in the central nervous system (CNS) by which oligodendrocyte (OLG) cell damage and death occurs. OLG produce myelin sheaths that surround neural axons, favoring nerve impulse propagation. Myelin sheaths coating greatly increases axonal conduction velocity, so its loss or degeneration causes neuronal axons demyelination, which is associated to sensory pathways nerve conduction velocity delays. It is for this reason that the sensorial alterations that occur during old age are especially important, not only because of their high prevalence, but also because of the direct repercussions that are generated on living organisms. It is therefore of utmost importance to evaluate the relationship between age and oxidative increase and its effect on different neuronal generators of sensory pathways in a primate model.

Keywords: brain electric activity, oxidative stress, evocated potentials, aging, *Macaca mulatta*

Introduction

The changes in the population pyramid towards the aging of the population have become a worldwide relevant topic due to its impact, not only in human health, but also at social, political and economic levels (Reques-Velasco 2006). The number and proportion of the elderly population has increased with respect to the other age groups duo to a persistent trend of fertility decline and an increase in health care, in a process that could eventually lead to an enduringly aged global population. It is an indisputable fact that 21st century societies will be noticeably older than any other time in human history. However, increased life span does not always come together with health-span, and in countless cases the last 10-12 years of a person's life are associated with degenerative diseases.

This concern has generated a number of interrogations about the medical and social services that will be required to attend to this population group (Ham-Chande 2011; WHO 2002). Therefore, trying to prevent age related diseases have become a matter of principal importance. So the main biological challenge is the understanding of the cellular and molecular mechanisms responsible for the functional deterioration occurring during aging and their possible prevention.

In the case of brain deterioration, we still do not know the molecular mechanisms that condition this phenomenon. There are numerous studies that have described the physiological changes generated by age in non-pathological conditions, such as neuronal degeneration, demyelination and deficits in cognitive functions, and have pointed towards oxidative stress as part of the decline process. Oxidative stress increase has also been related with the pathological outcomes that occur during neurodegenerative diseases such as Alzheimer's, Parkinson's and multiple sclerosis, among others (Hirrlinger et al., 2002; Sims-Robinson et al., 2013; Ibáñez- Contreras et al., 2014).

It is important to distinguish between aging as a universal physiological stage of life (which is influenced by multiple genetic and environmental factors) and disease (WHO 2002; González and Ham-Chande, 2007). However, it is now accepted that the aging process is a risk factor for sickness and disability, even though aging are disease are not synonyms.

During the last decades, a simple measure has been sought to determine oxidative stress in humans in clinical situations, nevertheless, to date, there is not a definite or gold standard method to measure systemic oxidative stress that could be directly applied to the human clinic (McTigue and Tripathi 2008). Usually the determinations are made in serum or plasma, where oxidation products derived from free radical-mediated reactions are used as oxidative stress indicators (Königsberg 2008, 2013).

Oxidative damage to macromolecules can be determined by different techniques. The most widely used markers for lipid oxidation are malondialdehyde (MDA), 4 hydroxynonenal (4HNE) and isoprostanes; for nucleic acids the golden standard is the adduct 8-hydroxy-2′-deoxyguanosine (8OHdG) and for proteins carbonylation, nitration and glycation are commonly used. Another indirect evidence of cellular oxidative stress is the increased expression and activity of molecules involved in the antioxidant defense, such as heme-oxygenases, superoxide dismutases (SOD), glutathione transferases, catalases, peroxidases and glucose-6-phosphate dehydrogenase, which are activated to counteract oxidants (Perry et al., 2012). However, it is not enough to measure a single oxidative stress marker, but the general picture must be analyzed in order to understand the overall response, in particular during aging. A new tool that can contribute to the whole process evaluation are the evoked potentials (EP's), which provide detailed measurements of neural sensory processing over time, thus improving to the global analysis of the nature and type of neuronal changes associate to age (Ibáñez-Contreras et al., 2016). Nevertheless, there are still not enough studies where changes in the brain redox state are related to the neurophysiological changes associated to increasing age.

Brain Aging

Aging can be defined as a stage with an important decline in the physiological functions. This stage involves biological, psychological and socio-cultural factors that all tougher increase the prevalence chronic-degenerative diseases (Poon et al., 2004; Díaz-Piñera et al., 2010; Sala-Llonch et al., 2014). The World Health Organization (WHO) refers to aging as a continuous, heterogeneous, universal, and irreversible process that leads to a progressive loss of adaptive capacity (WHO 2002; Arcos-Correa and González-Martínez 2009), generated from morphological, physiological and metabolic changes that occur with age and do not result from disease or extrinsic agents (WHO, 2002). Although aging is not a disease or a pathology by itself, most of the definitions that try to describe the aging process, include both, the physiological stage and the pathological process, so it is often debated whether aging and disease constitute a dichotomy or simply a continuum of the same phenomenon (Tingo et al., 2004).

Increased life expectancy in humans has augmented the occurrence of neurological deficit and degenerative changes. These changes do not always indicate neuronal death but they have been associated to decreased capacity to adapt to difficult conditions, such as oxidative stress, along with a motor and sensory dysfunction that generally are accompanied by a progressive functional deterioration (Hirrlinger et al., 2002; Pichardo- Fuster and Pedrero-Nieto 2009). Thus, brain aging in turn can be defined as the loss of the ability to adapt to changes that might be necessary to perform the normal brain function (Escobar-Izquierdo 2001). This has been related to the alteration of the homeostatic oxidative balance and the way in which the organism responds to it (Hirrlinger et al., 2002, Königsberg 2008, Pérez et al., 2009, Sims-Robinson et al., 2013).

The free radical theory of aging is based on free radicals excessive production and increased accumulation of oxidative damage to biomolecules during organismal lifespan, which decrease biological functions and are associated to functional deterioration during old age (Harman et al., 1956). However, it has recently been shown that it is not enough to increase the antioxidant systems and decrease oxidative damage in order to delay aging, rather it has been described that the way organisms respond to stress is determinant (Pérez et al., 2009).

Oxidative Stress and Aging Effects on Nerve Conduction Velocity

It has been shown that with increased age there is selective neuronal atrophy, which leads to a progressive decrease of the dendritic tree and synapses number, mainly in the phylogenetically most recent cerebral cortex: the prefrontal cortex and parietotemporal cortex. In this sense, morpho-functional changes in dendrites show detrimental effects, since they are the primary substrate for neuronal information processing and they play a vital role in neural circuits functional properties (Dickstein et al. 2007). There is a strong link between dendritic changes and the post-synaptic neurotransmitters effects of cholinergic, serotonergic, dopaminergic and glutaminergic systems, which may impair intra and inter-neuronal communication. The resulting inability of neurons to function effectively, leads to alterations in cortical-cortical signaling pathways, such as those connecting the superior temporal cortex and prefrontal cortex. In particular, it has been demonstrated in humans and in NHP that the number of neurons expressing certain ionotropic glutamate receptors (GluR) and N-methyl D-aspartate (NMDA-R) receptors are significantly reduced during aging (Hof and Cols 2002; Dickstein et al. 2007).

It is well known that nerve conduction velocity is related to the particular nerve diameter and to its normal myelination status. Myelin allows the conduction of nerve impulses through the CNS, being the OLG the responsible cells for neurons axonal coating. At the peripheral level, the Schwann cells are the responsible ones for the myelin formation and for the peripheral nerve fibers coating. Therefore, demyelination of either nerve fibers or cerebral cortex nuclei, caused by the oxidative stress increase associated with age, leads to a decrease in nerve conductivity (Ibáñez-Contreras et al., 2014b, 2016).

During the last decades, it has been evidenced that with the increase of age, different deleterious mechanisms in the CNS are generated by which OLG damage and cell death occurs; among which are mitochondrial dysfunction, axonal transport deterioration, trophic support decrease, oxidable substrates increase, antioxidant enzymes low production, calcium homeostasis alterations, microglial activation increase, etc. (Smith et al., 1999, Dewar et al., 2003, Sims-Robinson et al., 2013; Di Penta et al., 2013). As mentioned before, OLG interaction with neurons through myelin sheath coating favors the action potential propagation; OLG demise or degeneration causes neuronal axons demyelination and therefore delays nerve conduction velocity at the central level (von Bernhardi, 2005).

In the case of the PNS, the decrease in the nerve conduction velocity has also been observed and it has been related to degenerative diseases. These has mainly been attributed to inflammatory processes in the peripheral nerves myelin sheaths, together with an antioxidant enzymes lower concentration (Smith et al., 1999, Sims-Robinson et al., 2013, Ibáñez-Contreras et al., 2016), along with axonal damage or axonal degeneration caused by mitochondrial dysfunction, ATP depletion, ion channels redistribution among others (Cantú et al., 2007; di Penta et al., 2013).

It has been described that during aging there is a continuous synapses loss, even in the absence of degenerative pathologies, and a decrease in permeability and neuronal excitability (Watson et al., 2012). Moreover, when the neuronal plasma membrane components are oxidized, they can promote important neurophysiological changes, which potentially alter a large variety of intrinsic electrophysiological parameters (Watson et al., 2012). Therefore, it is important to use different diagnostic tools to characterize the early, but progressive, stages of the neurodegenerative pathologies, because it is difficult to differentiate the stable mild alterations during non-pathological aging from the alterations associated to disease. Therefore, some electrophysiological techniques, such as evoked potentials (EP's), can be used as a diagnostic indicator of great sensitivity and specificity in the evaluation of sensory pathways.

Evoked Potentials in Sensory Pathways (EP's)

The structural and functional organization of living beings depends on communication systems that provide permanent information to the cells and organs, and regulate their activities, from the molecular level up to complex behaviors manifestations (Ibáñez-Contreras et al., 2016). Traditionally, sensory modalities such as vision, audition, tact, pain and chemical senses, among others, have been investigated in isolation. It is well known that in the nervous system (NS), the information recording, coding and analysis is performed on structures extending from the periphery to the spinal cord, brain stem, thalamus and cerebral cortex, what constitutes the brain sensorial organization (López-

Antúnez, 2003). Recently, the number of studies evaluating sensory intermodal integration has increased, demonstrating that multisensory integration provides clear advantages in terms of individuals' adaptation and survival (Lanz et al., 2013). It is for this reason that the sensorial alterations that occur during old age are especially important, not only because of their high prevalence, but also for the direct repercussions generated on living organisms (Sala-Llonch et al., 2014, Ibáñez-Contreras et al., 2016).

The term evoked potential (EV) can be defined as the average of multiple electrical responses or as the electrical record given after a single stimulus, by which peaks or waves were obtained, which are defined as positive or negative deflections. Each of them presents a specific latency, amplitude and polarity, and all together constitute the neural generators, defined in turn, as the synaptic transmission or the electric charge movement generated by the action potentials propagation in the nerve pathway (Misulis 1994; Fitzgerald 2012).

The EP's are records of the auditory, visual, somatosensory and motor pathways specific activities, which allow the evaluation of cortico-subcortical responses, produced by sensorial peripheral stimuli (Poblano, 2003). In addition to assessing the sensory pathway integrity, EP's can indicate if there is a problem in nerve conduction, characterize its severity and refer the lesion to a specific segment (Hernández-Godínez et al., 2011, 2012).

EP's General Description

As mentioned before, EP's consist of peaks and waves series that are generated from a particular stimulus. Each wave has the following characteristics (Figure 1):

- *Positive or negative polarity*: Polarity refers to the positivity and/or negativity between the two electrodes connected to the recording system inputs. Peaks can be named by a combination of the letters P or N, with reference to the positive or negative polarity of the peak, followed by a number representing the latency of the expected peak.
- *Sequential wave number*: Most EP's have more than one wave or peak, and this depends on the sensory or motor pathway being stimulated.
- *Latency*: Latency can be defined as the time in the electrical response generated by the body, starting from the stimulation of a certain sensory or motor pathway, which is measured in milliseconds (msec). Latency

varies depending on the evaluated pathway, age, sex and electrical recording systems.

- *Amplitude*: From a clinical point of view, the amplitude is a less useful measure than the latency. It represents the voltage difference between a peak and the reference level that has zero amplitude.
- *Wave morphology*: It is composed of different components or peaks with a variable latency and amplitude dependent on the stimulated sensory pathway.

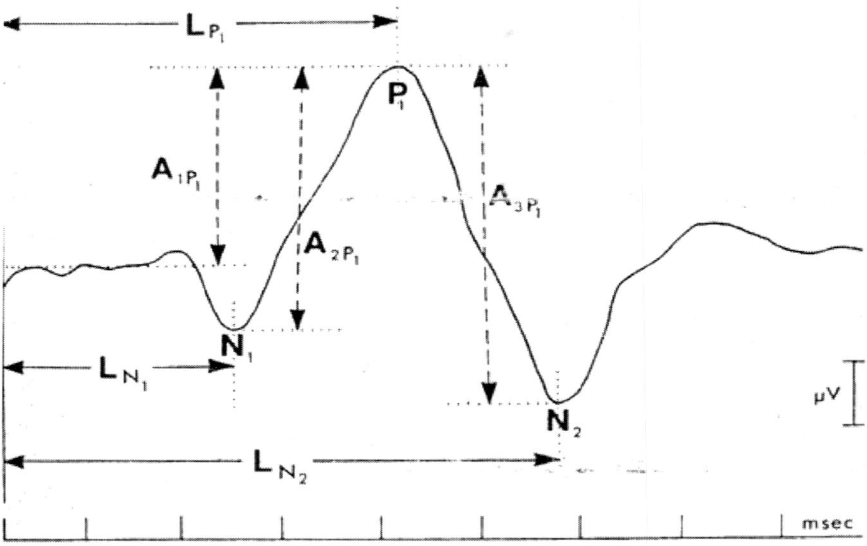

Figure 1. EP' parameters measurement. Latency peaks are measured from the first negative deflection after the stimulus (L_{N1}), followed by the first positive wave or peak (L_{P1}) and the second negative peak (L_{N2}). The wave amplitude is measured from P1 to a pre-responsible baseline (A_{1p1}), peak to peak from the previous polarity (A_{2P1}), or following the same peak-to-peak polarity (A_{3P1}) (Taken from Misulis 1994).

Based on the above, the recording parameters of each sensory pathway (auditory, visual, somatosensory) are different, showing changes in the quantity and morphology of the obtained waves, as well as in the latencies and amplitudes values. There are several EP's types, depending on the sensory pathway studied:

a) Brainstem Auditory Evoked Potentials (BAEP's)

BAEP's are short latency EP's produced by a brief stimulus, click or rarefaction clicks. They are characterized by at least four permanent positive waves in animals and five in humans, which reflect the electrical activity of the specific nuclei located in the brainstem (Figure 2) (Ibáñez-Contreras et al., 2011b, 2014a, 2016). The BAEP's are of great utility in various clinical situations, including coma, brain death, degenerative disorders, demyelinating diseases, infectious processes, metabolic disorders, tumors, etc. In neurological disorders evaluation, the diagnosis is typically focused on two categories that show the damage topographic location. The first one involves the VIII cranial nerve and the brain stem caudal position and the second one includes lesions affecting the brain stem rostral portion (Elías, 1997). Its use has also been described in demyelinating diseases based on waves latency and amplitude variable anomalies (Chiappa, 1980; Elías, 1997).

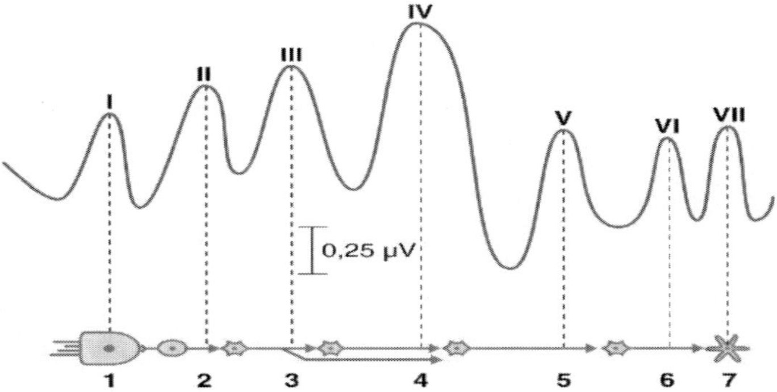

Figure 2. Brainstem Auditory Evoked Potentials (BAEP's).

Evoked potentials sources: 1, cochlear nerve distal portion (cochlear hair cells); 2, cochlear nerve (proximal); 3, cochlear nucleus; 4, lateral lemniscus; 5, lower coliculo (lower arm); 6, medial geniculate body (auditory radiation); 7, primary auditory cortex (Taken from Fitzgerald 2012).

b) Visual Evoked Potentials (VEP's)

VEP's allow visual pathway electrical activity evaluation from specific brain areas. The stimulation is performed with light flashes (stroboscopic light) or images of alternating pictures (checkers). VEP's are characterized by presenting at least 2 negative waves and 1 permanent positive wave in animals and in humans (Figure 3) (Solis-Chavez et al., 2014; Chiappa, 1980). Although the VEP's are very useful in retinal, neural and optic chiasm diseases assessment, their use in the diagnosis and pathophysiological monitoring of neurodegenerative diseases, such as Parkinson's disease (PD), has also been described. In PD patients a significant increase in latencies, in particular of the P wave, are observed. The latencies tend to increase proportionately to the disease progression (Delwaide et al., 1980; Nightingale et al., 1986).

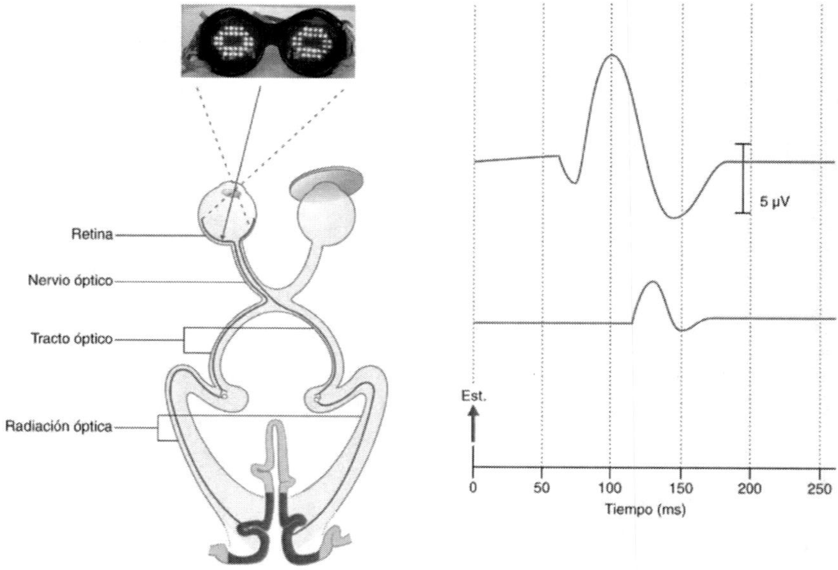

Figure 3. Visual evoked potentials obtained with luminous flashes stimulation. In VEP's registration, one eye is stimulated first and then the other. In the figure the anatomical structures of the visual pathway are shown, which represent the positive and negative deflections observed on the right side of the image, forming the evoked potential (Fitzgerald's modified image 2012).

c) Somatosensory Evoked Potentials (SSEP's)

SSEP's denote an objective measurement of the somatosensory pathway function and can be used to document and locate neuronal conduction interruption and delay. These types of studies are non-invasive diagnostic indicators that allow us to study the somatosensory pathways of both the CNS and the periphery (Figure 4) (Hernández-Godínez et al., 2011, 2012, 2014). SSEP's are used to evaluate peripheral sensory nerves damage in some pathologies and disorders such as metabolic disorders, including diabetes and uremia, multiple sclerosis, nutritional deficiencies, neurodegenerative diseases, axonal diseases including spinocerebral degeneration, nutritional disorders and polyneuropathies, among others (Pedersen and Trojaborg, 1982).

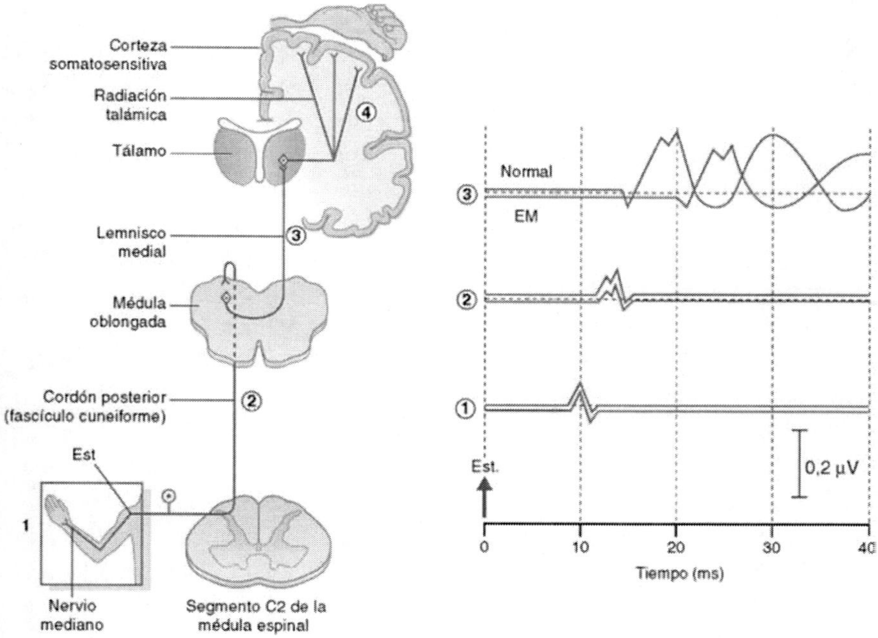

Figure 4. Short latency somatosensory evoked potentials derived after medial nerve stimulation. This pathway displays four segments: 1 is purely from the PNS; 2 is from the PNS from the brachial plexus to the spinal cord and from the CNS into the spinal cord; 3 and 4 are purely from the CNS. In the records obtained from a multiple sclerosis (MS) patient, path 1 is normal; Trace 2 has a reduced amplitude of negative (upward) and positive peaks within the posterior cord; Plot 3 shows an amplitude slowing and reduction (Image taken from Fitzgerald 2012).

NON-HUMAN PRIMATES AS A MODEL TO STUDY BRAIN AGING

To date, there are a large number of studies on cerebral aging and its relationship with oxidative stress and neurodegenerative diseases. In this context, nonhuman primates (NHP) have been one of the main animal models to study brain aging and neurodegeneration, due to their phylogenetic closeness with man (Nagahara et al., 2010; Kemnitz, 2011; Toledano et al., 2012; Ibáñez-Contreras et al., 2014 [a,b], 2016). The great diversity of existing primate species provides a unique opportunity for comparative studies of brain organization from a phylogenetic and ontogenic point of view. In this sense, sensory organization comparative studies have provided important advances in the understanding of brain-behavior relationships in a wide variety of primate species by comparing the cerebral cortex specific characteristics and their sensory, motor or cognitive functions (Carlson and Nystrom, 1994; Kaas, 2008). The NHP most studied species related to the neuropathological aspects correspond to the families *Cherirogaleidae* and *Cercophitecoideae* (Toledano et al., 2012). Within the family *Cercophitecoideae* the species most used in biomedical research are green monkeys (*Chlorocebus aethiops*), rhesus monkeys (*Macaca mulatta*) and tailed macaque monkeys (Kemnitz et al., 2000, Hernández- Godínez et al., 2011, 2012; Ibáñez-Contreras et al., 2011 [a,b]). It is estimated that the life span of these animals is approximately 30 years under captive conditions. Individuals who are over 20 years of age are described as senile and those who reach the age of 30 are defined as geriatric (Fowler et al., 2010; Ibáñez-Contreras et al., 2011[a,b]; Toledano et al., 2012).

It should be noted that within the primate order, it has been established that only humans present a significant degree of pathological changes in the brain, than have been associated with cognitive, functional and anatomical deficits, in which the demyelinating processes stand out. In the case of NHP, pathological alterations in the brain show a less consistent functional impairment than in humans, this is why NHP offer an extraordinary opportunity to study brain aging in the absence of the cellular changes that appear in various neurodegenerative diseases (González-Fraguela et al., 2000, Hof et al., 2002, Qin et al., 2006, Willette et al., 2010, Kemnitz 2011, Castro et al., 2012, Csiszar et al., 2012, Didier et al., 2016, Ibáñez-Contreras et al., 2014[b], 2016).

As previously mentioned, age, oxidative stress and demyelinating processes are important factors in neurophysiological records. A reduction of 60% in catecholamines biosynthesis has been described in rhesus monkeys, mainly in the sensory and association areas, which have been related to cognitive deficiencies, neurodegenerative diseases and neurotransmitter-mediated cerebral electrical activity delay (Grondin et al., 2003). The above is associated with what has already been said, when the myelin is destroyed or damaged, the nerve fibers are affected and therefore the ability of the nerves to conduct the electrical impulse to and from the brain is lost (Königsberg 2008; Watson et al., 2012; Ibáñez-Contreras et al., 2011, 2014[a,b], 2016).

In addition to the above, different NHP species have been used in diverse aging studies, like antioxidant enzymes levels determinations in senile monkeys (González-Fraguela et al., 2000) and caloric restriction (RC) studies, in species such as *Saimiri sciureus* and *Macaca mulatta*, where different oxidative stress indicators were evaluated (Kemnitz et al., 1994; Cols, 1997; Qin et al., 2006; Willette et al., 2010; Fowler et al., 2010). Research has also been done in order to understand how aging-associated oxidative stress contributes to liver tissue damage and disease in rhesus monkeys, with a direct correlation between mitochondrial DNA (mtDNA) injury and age-related injury. Greater lipid and protein oxidation was observed in older monkeys, as well as a decrease in antioxidant enzymes. Overall, these results suggest that the aging process is associated with defective mitochondria that cannot be efficiently eliminated. Damaged mitochondria produce large amounts of ROS harming mtDNA and other biomolecules (Castro et al., 2011). There are other studies that have reported an increased antioxidant enzymatic activity with age, however this increment is not enough to compensate for the oxidative damage (Ibáñez-Contreras et al., submitted)

Additionally, it has been described that some geriatric NHP develop abnormalities, both behavioral and morphological, similar to what occurs in senile humans (Bachevalier et al., 1991, Prince et al., 1991, Gearing et al., 1994, Tumminja et al., 1996, Brownell et al., 1998, Sani et al., 2003, Duce et al., 2008, Valenzuela et al., 2010, Ohnuki et al., 2010, Toledano et al., 2012, Bendle et al., 2011, Yue et al., 2012, Csiszar et al., 2012). This is why Parkinson disease models have been generated using NHP models. Different neurotoxic agents have been used to induce neuronal damage to the nigrostriatal pathway dopaminergic bodies (Tabbal et al., 2012). One of the most widely used is 1-methyl-4-phenyl-1,2,3,6-tetrahydropyridinium (MPTP), which induces selective degeneration to dopaminergic neurons in the substantia nigra and provokes behavioral alterations similar to those observed in humans with PD, including spontaneous activity reduction, stiffness, bradykinesia and tremor. Parkinson's symptoms are initially transient in NHP, but become permanent with repeated administration of the toxin, thus showing histopathological and neurochemical characteristics similar to those observed in human patients with this pathology (Blesa et al., 2010; Alarcón- Aguilar et al., 2010).

Our group and others have observed a progressive audition reduction in NHP as they age (Torre, 2000; Torre, 2001; Fowler et al., 2002; Fowler et al., 2010; Ibáñez-Contreras et al., 2011, 2014, 2016). Hearing loss reaches such a low degree that in geriatric monkeys, it becomes impossible to identify some components of the auditory pathway, which have been related to some type of cochlear or retrocochlear lesion produced by age-specific degenerative changes. It has also been observed that geriatric monkeys tend to increase the neuronal components absolute latencies of the auditory pathway. This has been related to hormonal changes, mainly to the decrease in the steroid hormones production during aging, as observed in humans and other animals (Hultcrantz et al., 2006; Fowler et al., 2010; Ibáñez-Contreras et al., 2011, 2014[a], 2016) (Figure 5).

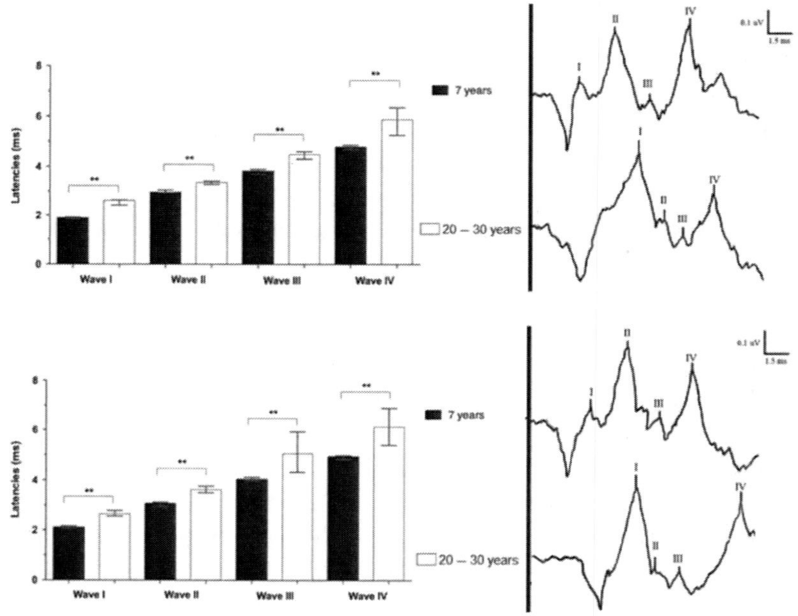

Figure 5. Brainstem Auditory Evoked Potentials (BAEP's) in rhesus monkeys (*Macaca mulatta*). Auditory pathway waves statistical analysis and morphology were obtained through BAEP's. The two-tailed Mann-Whitney U test showed statistically significant differences in all auditory pathway neural generators between the geriatric monkey group versus the control group. ** $p < 0.001$. (Ibáñez-Contreras et al., 2016).

According to Solís-Chavez and coworkers (2014), there is an evident increase in geriatric rhesus monkey VEP's latencies, which has been mainly related to a myelin defective regeneration and to the retinal ganglion cells degeneration. Optic nerve, anterior commissure and corpus callosum splenius morphological analyzes in old monkeys have evidenced some transverse profiles of myelinated axons with dense axoplasm, indicating sensory pathway degeneration. The frequency of these axonal degeneration profiles increases with age, and the result is more than 40% of anterior commissure nerve fibers loss at the optic nerve, as well as 25% of splenium nerve fibers loss (Sandell and Peters 2003, Bowley et al., 2010). Hence, it could be inferred that those neurophysiological alterations might be producing the visual pathway neural generators latency delay shown in Figure 6.

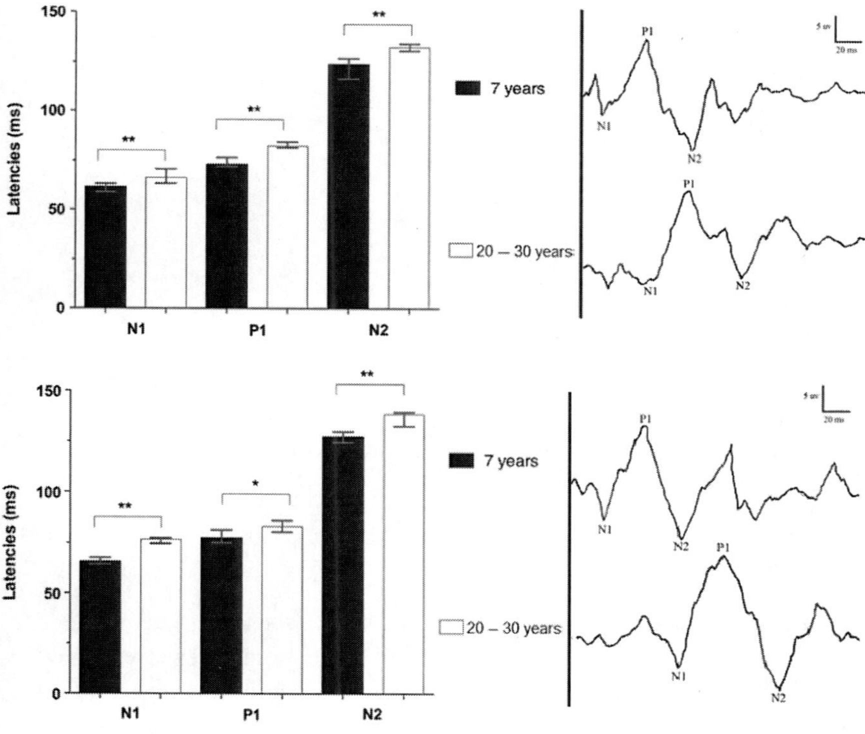

Figure 6. Visual evoked potentials (VEP's) in rhesus monkeys (*Macaca mulatta*). Visual pathway waves morphology and statistical analysis were obtained through VEP's. The two-tailed Mann-Whitney U test revealed statistically significant differences in all neural pathway generators among the geriatric monkey group versus the control group. ** p <0.001; * p <0.01 (Ibáñez-Contreras et al., 2016).

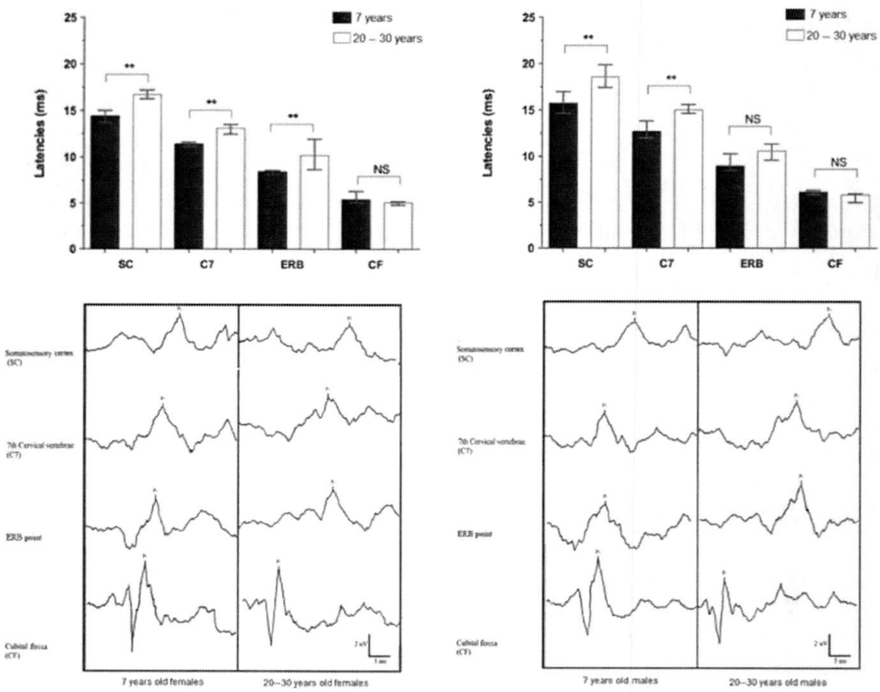

Figure 7. Medial nerve somatosensory evoked potentials (MNSSEP) in rhesus monkeys (*Macaca mulatta*). Anterior limbs pathway waves morphology and statistical analysis were obtained through MNSSEP. The two-tailed Mann-Whitney U test revealed statistically significant differences between the geriatric monkeys group versus the control group. ** $p < 0.001$; * $p < 0.01$. NS: Not significant (Ibáñez-Contreras et al., 2016).

SSEP's have been used to distinguish somatosensory system integrity from the peripheral nerves to the cortex. The somatosensory pathway crosses the spinal cord, posterior column, gracilis and cuneiform nucleus, medial lemniscus, thalamic nuclei ventrobasales, thalamo-cortical pathways, ending in the primary sensory cortex (Choi et al., 2001). It has been described in primates that brain white matter integrity diminishes with advanced age, being that decrease more pronounced in the anterior cerebral regions (Bowley et al., 2010; Coxon et al., 2012). Electro-microscopic analysis in rhesus monkeys have shown compacted myelin sheaths associated to degenerative alterations (Bowley et al., 2010). In addition, there is a general loss of myelinated nerve fibers, as well as concomitant changes occurring in gray matter and dopaminergic neurotransmission (Coxon et al., 2012). As shown in Figure 7, nerve conduction velocity in the older monkeys is longer than in the rest of the groups. This is in agreement with what has been described by Kalisch et al., (2010), where nerve conduction velocity decrease and sensory action potential nerve activation are related to a nerve fibers density decline, as well as a to a decrease in myelin thickness in the remaining fibers. These authors suggest that aging in combination with peripheral nervous system (SNP) and central nervous system (CNS) alterations may drastically affect somatosensory information processing; for example, tactile acuity decreases in old age. Tactile acuity reduction negatively affects fine motor activity performance, indicating that there is a greater need for tactile information in the elderly to achieve correct sensorimotor functioning (Kalisch et al., 2010).

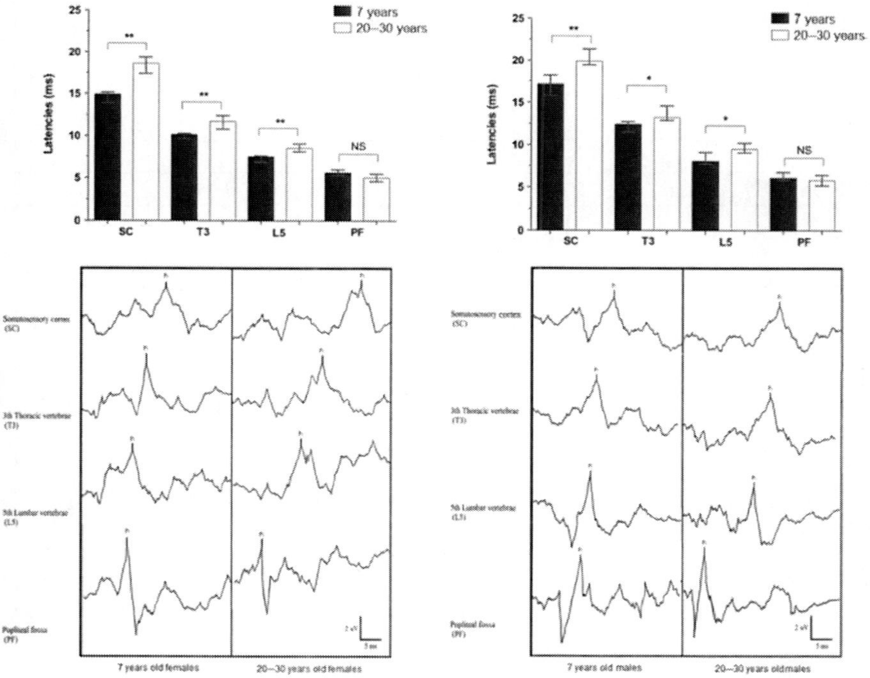

Figure 8. Tibial nerve somatosensory evoked potentials (TNSSEP) in rhesus monkeys (*Macaca mulatta*). Posterior limbs pathway waves morphology and statistical analysis were obtained through TNSSEP. The two-tailed Mann-Whitney U test revealed statistically significant differences between the geriatric monkeys group versus the control group. ** p <0.001; * p <0.01. NS: Not significant (Ibáñez-Contreras et al., 2016).

Table 1. Absolute latency values obtained from sensory pathways neural generators in rhesus monkeys (*Macaca mulatta*) (Ibáñez-Contreras et al., 2016)

Neural Generators	Female latency values (ms)		Male latency values (ms)	
BAEP's	7 years Ẋ ± SD	>20 years Ẋ ± SD	7 years Ẋ ± SD	>20 years Ẋ ± SD
Wave I	1.88 ± 0.03	2.56 ± 0.12	2.11 ± 0.05	2.69 ± 0.22
Wave II	2.99 ± 0.06	3.35 ± 0.09	3.05 ± 0.09	3.61 ± 0.18
Wave III	3.78 ± 0.13	4.44 ± 0.21	4.05 ± 0.07	5.16 ± 0.89
Wave IV	4.77 ± 0.09	5.82 ± 0.56	4.91 ± 0.08	6.16 ± 0.80
VEP's				
N1	60.40 ± 3.22	66.99 ± 3.57	65.14 ± 2.28	76.17 ± 1.63
P1	73.72 ± 2.37	82.78 ± 1.53	77.59 ± 3.34	83.17 ± 2.71
N2	120.62 ± 1.08	132.59 ± 2.08	127.48 ± 2.86	136.77 ± 3.50
MNSSEP				
SC	14.30 ± 0.84	16.71 ± 0.48	15.72 ± 1.21	18.61 ± 1.21
C7	11.35 ± 0.28	13.03 ± 0.57	12.83 ± 0.87	15.01 ± 0.58
ERB	8.36 ± 0.22	10.27 ± 1.58	9.28 ± 1.03	10.49 ± 0.90
CF	5.58 ± 0.66	5.02 ± 0.23	6.09 ± 0.30	5.55 ± 0.56
TNSSEP				
SC	14.58 ± 0.82	18.43 ± 1.06	16.81 ± 1.69	20.93 ± 1.54
T3	9.90 ± 0.59	11.67 ± 0.85	12.17 ± 0.86	14.31 ±1.51
L5	7.23 ± 0.63	8.61 ± 0.58	8.38 ± 0.87	10.07 ± 1.10
PF	5.86 ± 0.38	5.09 ± 0.52	6.22 ± 0.45	5.64 ± 0.72

Ẋ, mean; SD, standard deviation; CS, somatosensory cortex; 7th cervical vertebrae; ERB, Erb's Point, CF, Cubital Fossa; SC, somatosensory cortex; T3, 3rd thoracic vertebrae, L5, 5th lumbar vertebrae; PH, Popitleal Fossa

CONCLUSION

NHP use as a biological model in the biomedical and public health sciences has been a core element to significantly boost research. Given the difficulties in the evaluation and study of normal aging in humans, a series of investigations have been carried out using non-human primates to facilitate the evaluation of the changes generated during normal aging, independently of neurodegenerative alterations and post mortem artifacts. The EP's have been a very important support to evaluate the sensory pathways functionality and integrity, especially in non-collaborating patients like rhesus monkeys. The standardization of diverse techniques using this animal model within the neurophysiological sciences, has collaborated in the understanding of the demyelination process, as well as the metabolic changes that occur naturally during the aging process.

ACKNOWLEDGMENTS

This work was supported by CONACyT's grant FON.INST/298/2016. Ibáñez-Contreras, Hernández-Godínez, and Hernández-Arciga are CONACyT scholarship holders.

REFERENCES

Alarcón-Aguilar, A., Santamaría del Ángel, A., Königsberg-Fainstein, M., (2010). Modelos neurotóxicos de la enfermedad de Parkinson y disfunción mitocondrial. *REB*. 29, 92-100.

Arcos-Correa J.H., González–Martínez J.F., (2009) Capítulo 1. Determinantes de la longevidad humana. En: Geriatría. Eds. González-Martínez J.Fco., Pichardo-Fuster A, García L. Ed. Mc Graw Hill, México D.F. p 1-18.

Bachevalier, J., Landis, L.S., Walker, L.C., Brickson, M., Mishkin, M., Price, D.L., Cork, L.C., (1991). Aged monkeys exhibit behavioral deficits indicative of widespread cerebral dysfunction. *Neurobiol. Aging*. 12, 99-111.

Blesa, J., Juri, C., Collantes, M., Peñuelas, I., Prieto, E., Iglesias, E., Martí-Climent, J., Arbizu, J., Zubieta, J.L., Rodríguez-Oroz, M.C., García-García, D., Richter, J.A., Cavada, C., Obeso, J.A., (2010). Progression of dopaminergic depletion in a model of MPTP-induced Parkinsonism in non-human primates. An (18) F-DOPA and (11) C-DTBZ PET study. *Neurobiol. Dis*. 38, 456:463.

Bowley, M.P., Cabral, H., Rosene, D.L., Peters, A., (2010). Age changes in myelinated nerve fibers of the cingulated bundle and corpus callosum in the rhesus monkey. *J. Comp. Neurol.* 518, 3046-3064.

Brownell, A.L., Jenkins, B.G., Elmaleh, D.R., Deacon, T.W., Spealman, R.D., Isacson, O., (1998). Combined PET/MRS brain studies show dynamic and long-term physiological changes in a primate model of Parkinson disease. *Nat. Med.* 4, 1308-1312.

Cantú- Brito, C., Baizábal, C.F., Aldrete, V.J., Chávez, G.E., (2007). Aplicaciones clínicas del acido tióctico (ácido α-lipoico) en pacientes con neuropatía diabética sintomática: revisión. *Med. Int. Mex.* 23, 151-158.

Carlson, M., Nystrom, P., (1994). Tactile discrimination capacity in relation to size and organization of somatic sensory cortex in primates: I. Old-world Prosimian, Galago; II. New-world Anthropoids, Saimiri and Cebus. *J. Neurosci.* 14, 1516-1541.

Castro, M.R., Suarez, E., Kraiselburd, E., Isidro, A., Paz, J., Ferder, L., Ayala-Torres, S., (2012). Aging increases mitocondrial DNA damage and oxidative stress in liver of Rhesus monkeys. *Exp. Gerontol.* 47, 29-37.

Chia, L.S., Thompson, J.E., Moscarello, M.A., (1983). Changes in lipid phase behavior in human myelin during maturation and aging. *FEBS J.* 157, 151-158.

Chiappa, K.H., (1980). Brainstem auditory evoked responses in 200 patients with multiple sclerosis. *Ann. Neurol.* 7, 135-143.

Choi, Y.C., Kim, Y.D., Kim, W.J., Yang, J.W., Moon, J.S., (2001). The central conduction time in posterior tibial and pudendal nerve somatosensory evoked potentials. *Yonsei. Med. J.* 42, 9-13.

Coxon, J.P., Van Impe, A., Wenderoth, N., Swinnen, S.P., (2012) Aging and inhibitory control of action: cortico-subthalamic connection strength predicts stopping performance. *J. Neurosci.* 32, 8401-8412.

Csiszar, A., Podlutsky, A., Podlutskaya, N., Sonntag, W.E., Merlin, S.Z., Philipp, E.E.R., Doyle, K., Davila, A., Recchia, Fa., Ballabh, P., Pinto, J.T., Ungvari, Z., (2012). Testing the oxidative stress hypothesis of aging in primate fibroblasts: Is there a correlation between species longevity and cellular ROS production?. *J. Gerontol. A. Biol. Sci. Med. Sci.* 67, 841-852.

Delwaide, P.J., Mesraoua, B., De Pasqua, V., (1980). Visual evoked potentials in Parkinson's disease. *Rev. Electroencephalogr. Neurophysiol. Clin.* 10, 338-342.

Dewar, D., Underhill, S.M., Goldberg, M.P., (2003). Oligodendrocytes and ischemic brain injury. *J. Cereb. Blood. Flow. Metab.* 23, 263-274.

Díaz-Piñera, W.J., García-Villar, Y., Linares Fernández, T.M., Rabelo Padua, G., Díaz Padrón, H., (2010). Envejecimiento e invalidez. Nuevos retos para la sociedad cubana. *Revista cubana de salud y trabajo.* 11: 38-48.

Didier, E.S., MacLean, A.G., Mohan, M., Didier, P.J., Lackner, A.A., Kuroda. M.J., (2016). Contributions of nonhuman primates to Research on aging. *Vet. Pathol.* 53, 277-290.

Dickstein, D.L., Kabaso, D., Rocher, A.B., Luebke, J.I., Wearne, S.L., Hof, P.R., (2007). Changes in the structural complexity of aged brain. *Aging cell.* 6, 275-284.

Di Penta, A., Moreno, B., Reix, S., Fernandez-Diez, B., Villanueva, M., Errera, O., Escala, N., Vandenbroeck, K., Comella, J.X., Villoslada, P., (2013). Oxidative stress and proinflammatory cytokines contribute to demyelinaton and axonal damage in a cerebellar cultura model of neuroinflammation. *PLoS One*. 8, e54722.

Duce, J.A., Podvin, S., Hollander, W., Kipling, D., Rosene, D.L., Abraham, C.R., (2008). Gene profile analysis implicates Klotho as an important contributor to aging changes in brain white matter of the rhesus monkey. *Glia* 56,106-117.

Elías Y. (1997) Potenciales provocados auditivos del tallo cerebral en el diagnóstico del daño auditivo y neurológico del adulto. En: Hernández OA, Flores R T, Peñaloza LY (eds). Registros electrofisiológicos para el diagnóstico de la patología de la comunicación humana. Instituto de la comunicación humana. México. SSA, p. 157-170.

Escobar-Izquierdo A., (2001). Envejecimiento Cerebral Normal. *Rev. Mex. Neuro.* 2, 197- 202.

Fitzgerald MJT, Gruener G, Mtui E. Neuroanatomia clinica y neurociencia. Capitulo 31. Potenciales Evocados. Elsevier Saunders. 6° Edición. 2012Pp 323, 330.

Fowler, C.G., Chiasson, K.B., Leslie, T.H., Thomas, D., Beasley, T.M., Kemnitz, J.W., Weindruch, R., (2010). Auditory function in rhesus monkeys: effects of aging and caloric restriction in the Wisconsin monkeys five years later. *Hear Res.* 261, 75-81.

Fowler, C.G., Torre, P. 3rd, Kemnitz, J.W., (2002). Effects of caloric restriction and aging on the auditory function of rhesus monkeys (Macaca mulatta): The University of Wisconsin Study. *Hear Res.* 169, 24-35.

Gearing, M., Rebeck, G.W., Hyman, B.T., Tigges, J., Mirra, S.S., (1994). Neuropathology and apolipoprotein E profile of aging chimpanzees: implications for Alzheimer disease. *Proc. Natl. Acad. Sci. USA.*, 91, 9382-9386.

González-Fraguela, M.E., Robinson, M.A., Delgado, R., (2000). Estudio de indicadores de estrés oxidativo en monos viejos. *Rev. Neurol.* 31, 1012-1015.

González, C.A. y Ham-Chande, R., (2007). Funcionalidad y salud: una tipología del envejecimiento en México. *Salud Publica Mex.* 49, S448-s458.

Grondin, R., Cass, W.A., Zhang, Z., Stanford, J.A., Gash, D.M., Gerhardt, G.A., (2003). Glial cell line-derived neurotrophic factor increases stimulus-evoked dopamine release and motor speed in aged rhesus monkeys. *J. Neurosci.* 23, 1974-1980.

Ham-Chande, R., (2011). Diagnóstico socio-demográfico del envejecimiento en México. CONSEJO NACIONAL DE POBLACIÓN, ISBN: 978-607-427-112-6.

Harman, D., (1956). Aging: a theory based on free radical and radiation chemistry. *J. Gerontol.* 17, 249-258.

Haynes, R.L., Folkerth, R.D., Szweda, L.I., Volpe, J.J., Kinney, H.C., (2006). Lipid peroxidation during human cerebral myelination. *J. Neuropathol. Exp. Neurol.* 65, 894-904.

Hernández-Godínez, B., Ibáñez-Contreras, A., Durand-Rivera, A., Reyes-Pantoja, S.A., Ramírez-Hernández, R., Rodríguez-Guzmán, P., Tena-Betancourt, E., (2011). Somatosensory Evoked Potentials SEPs, of Median and Tibial Nerves in Rhesus Monkeys (Macaca mulatta) Under Captivity: Influence of Ontogenic Status in Neonatal, Infant, Young, Adult and Senile Stages. *J. Med. Primatol.* 40, 79-87.

Hernández-Godínez, B., Ibáñez-Contreras, A., Reyes-Pantoja, S.A., Durand-Rivera, A., Galvan-Montaño, A., Perdigón-Castañeda, G., Climent- Plamer, Ma.C., Tena Betancourt, E., (2012). Preliminary Studies of Neurosensory and Cardiopulmonary Health Compromise in Captive– Bred Rhesus Macaques (Macaca mulatta) Suffering Scoliosis. *J. Med. Primatol.* 41, 163-171.

Hernández-Godínez, B., Bonilla-Jaime, H., Poblano, A., Arteaga-Silva, M., Bautista-Rodríguez, M.A., Solís-Chávez, S., Heras-Romero, Y., Tena-Betancourt, E., Ibáñez-Contreras, A., (2014). Ontogeny of Somatosensory Evoked Potentials of Median and Tibial Nerves in Rhesus Monkeys (Macaca mulatta): Influence of Dissociative Anesthetic Mixtures under Captivity Conditions. Capítulo 3. *En Monkeys: Brain development, Social and hormonal mechanisms and Zoonotic diseases*. Editorial Nova Science Publishers, Inc. USA, pp 51-78.

Hirrlinger, J., Resch, A., Gutterer, J.M., Dringen, R., (2002). Oligodendroglial cells in culture effectively dispose of exogenous hydrogen peroxide: comparison with cultured neurons, astroglial and microglial cells. *J. Neurochem.* 82, 635-644.

Hof, P.R., Gilissen, E.P., Sherwood, C.C., Duan, H., Lee, P.W.H., Delman, B.N., Naidich, P., Gannon, P.J., Perl, D., Erwin, J.M., (2002). Comparative Neuropathology of Brain Aging in Primates. Aging in Nonhuman Primates. Erwin JM, Hof PR (eds). *Interdiscipl Top Gerontol. Basel, Krager,* vol 31, pp 130-154.

Hultcrantz, M., Simonoska, R., Stenberg, A.E., (2006). Estrogen and hearing: a summary of recent investigations. *Acta Otolaryngol.* 126, 10-14.

Ibáñez-Contreras, A., Hernández-Godínez, B., Reyes-Pantoja, S.A., Jiménez-García, A., Morales-Cruz, E., Justo-Juárez, T.P., Galván-Montaño, A., (2011). Clinical Evaluation in geriatric rhesus monkeys (Macaca mulatta): Through serological studies and physiological constants in captivity. *JAVA*. 10, 3281-3286.

Ibáñez-Contreras, A., Durand-Rivera, A., Hernández-Godínez, B., Reyes-Pantoja, S.A., Morales-Martínez, J.J., (2011). Potenciales evocados auditivos de tallo cerebral en monos rhesus (Macaca mulatta) en diferentes etapas fisiológicas en condiciones de cautiverio. *Arch. Med. Vet.* 43: 223-232.

Ibáñez-Contreras, A., Solís-Chávez, S.A., Tena-Betancourt, E., Hernández-Godínez, B., (2014). Brainstem's auditory evoked potentials and intervals values in rhesus monkey (Macaca mulatta) at different intensities (70-30 dB). Capitulo 1. *En Monkeys: Brain development, Social and hormonal mechanisms and Zoonotic diseases.* Editorial Nova Science Publishers, Inc. USA, pp 3-34.

Ibáñez-Contreras, A., Poblano, A., Hernández-Godínez, B., Alarcón-Aguilar, A., Arteaga-Silva, M., Tena-Betancourt, E., Königsberg, M., (2014). The Use of Non-Human Primates As Bio-Models for Aging Brain Studies: Interventional Role of Oxidative Stress in Demyelization Processes. Capítulo 2. *En Monkeys: Brain development, Social and hormonal mechanisms and Zoonotic diseases.* Editorial Nova Science Publishers, Inc. USA, pp 35-50.

Ibáñez-Contreras, A., Poblano, A., Arteaga-Silva, M., Hernández-Godínez, B., Hernández-Arciga, U., Toledo, R., Konigsberg, M., (2016). Visual, auditive and somatosensory pathways alterations in geriatric rhesus monkeys (Macaca mulatta). *J. Med. Primatol.* 45, 92-102.

Kaas, J.H., (2008). The evolution of the complex sensory and motor systems of the human brain. *Brain. Res. Bull.* 75, 384-390.

Kalisch, T., Tegenthoff, M. and Dinse, H. R., (2010). Repetitive Electric Stimulation Elicits Enduring Improvement of Sensorimotor Performance in seniors. *Neural Plast.* 2010, 690531.

Kemnitz, J.W., Roecker, E.B., Weindruch, R., Elson, D.F., Baum, S.T., Bergman, R.N., (1994). Dietary restriction increases Insulin sensitivity and lowers blood glucose in rhesus monkeys. *Am. J. Physiol.* 266, E540-E547.

Kemnitz, J.W., Roecker, E.B., Haffa, A.L., Pinheiro, J., Kurzman, I., Ramsey, J.J., Mac Ewen, E.G., (2000). Serum dehydroepiandrosterone sulfate concentrations across the life span laboratory housed rhesus monkeys. *J Med Primatol.* 29, 330-337.

Kemnitz, J.W., (2011). Calorie restriction and aging in nonhuman primates. *ILAR J.* 52, 66-77.

Kim, M.J., Aiken, J.M., Havighurst, T., Hollander, J., Ripple, M.O., Weindruch, R., (1997). Adult-Onset Energy Restriction of Rhesus Monkeys Attenuates Oxidative Stress–Induced Cytokine Expression by Peripheral Blood Mononuclear Cells. *J Nutr.* 127, 2293-2301.

Königsberg, F.M., (2008). Radicales libres y estrés oxidativo. Aplicaciones médicas, 1a ed., México: El Manual Moderno México.

Königsberg F.M. (2013) Radicales libres y estrés oxidante. Capítulo 24. Bioquímica de Laguna. 7ª edición. Laguna J, Piña Garza E, Martínez Montes F, Pardo-Vazquez JP, Riveros Rosas H (eds). El Manual Moderno México.

Lanz, F., Moret, V., Rouiller, E.M., Loquet, G., (2013). Multisensory integration in non-human primates during a sensory-motor task. *Front Hum Neurosci.* 7, 1-15.

López –Antúnez, L., (2003). Anatomía Funcional del sistema nervioso. Noriega Editores. Limusa, México.

Marotti, T., Sobocanec, S., Macak-Safranko, Z., Saric, A., Kusic, B., Balog, T., (2010). Sensitivity to oxidative stress: sex matters. *Med Sci.* 35, 59-68.

McTigue, D.M., Tripathi, R.B., (2008). The life, death, and replacement of oligodendrocytes in the adult CNS. *J. Neurochem.* 107, 1-19.

Misulis, K.E. Spehlmann's Evoked Potential Primer. Visual, Auditory, and Somatosensory evoked potentials in clinical diagnosis. Segunda edición. Butterworth-Heinemann. USA.

Nagahara, A.H., Bernot, T., Tuszynski, M.H., (2010). Age-related cognitive deficits in rhesus monkeys mirror human deficits on an automated test battery. *Neurobiol Aging.* 31, 1020–1031.

Nightingale, S., Mitchell, K.W., Howe, J.W., (1986). Visual evoked cortical potentials and pattern electroretinograms in Parkinson's disease and control subjects. *J Neurol Neurosurg Psychiatry.* 49, 1280-1287.

Ohnuki, T., Nakamura, A., Okuyama, S., Nakamura, S., (2010). Gene expression profiling in progressively MPTP-lesioned macaques reveals molecular pathways associated with sporadic Parkinson's disease. *Brain Res.* 1346, 26-42.

Organización Mundial de la Salud (OMS). (2002). Envejecimiento Activo: un marco político. *Rev Esp Geriatr Gerontol.* 37, 74-105.

Pedersen, L., Trojaborg, W., (1981). Visual, auditory and somatosensory pathway involvement in hereditary cerebellar ataxia." *Electroencephalogr Clin Neurophysiol.* 52, 283-297.

Pérez, V.I., Van Remmen, H., Bokov, A., Epstein, C.J., Vijg, J., Richardson, A., (2009). The overexpresión of major antioxidant enzymes does not extend the lifespan of mice. *Aging Cell.* 8, 73-75.

Perry, G., Mondragón-Rodríguez, S., Nunomura, A., Zhu, X., Moreira, P.I., Smith, M.A., (2012). Capítulo 3: Estrés y equilibrio oxidativos en las enfermedades neurodegenerativas. En: *Neurodegeneración. Patología molecular de la demencia y los transtornos del movimiento. Dickson DW and Weller RO (Eds).* Segunda Edición. Editorial Médica Panamericana S.A., Madrid, España.

Pichardo- Fuster A. y Pedrero-Nieto L. (2009) Capítulo 5. El anciano frágil, cascada de la fragilidad, falla que medra y pérdida de la vitalidad (deslizamiento). En: *Geriatría. Eds. González-Martínez JFco.,* Pichardo-Fuster A, García L. Ed. Mc Graw Hill, México D.F. p 69-81.

Poblano A. (2003) Detección y Estimulación tempranas del Niño con Daño Neurológico. Instituto Nacional de Perinatología. Editores de Texto Mexicanos. México D.F.

Poon, H.F., Calabrese, V., Scapagnini, G., Butterfield, A.D., (2004). Free radicals and brain aging. *Clin Geriatr Med.* 20, 329-359.

Qin, W., Chachich, M., Lane, M., Roth, G., Bryant, M., de Cabo, R., Ottinger, M.A., Mattison, J., Ingram, D., Gandy, S., Pasinetti, G.M., (2006). Calorie restriction attenuates Alzheimer's disease type brain amyloidosis in Squirrel monkeys (Saimiri sciureus). *J Alzheimers Dis.* 10, 417-422.

Reques-Velasco P. (2006) Capítulo 3. El análisis demográfico del envejecimiento, conceptos, técnicas y métodos. En: *Biogerontología. Eds Damaso Crespo Santiago.* Ed. Universidad de Cantabria. España. p 55-76.

Sala-Llonch, R., Junqué, C., Arenaza-Urquijo, E.M., Vidal-Piñeiro, D., Valls-Pedet, C., Palacios, E.M., Domenech, S., Slva, A., Bargalló, N., Bartres-Faz, D., (2014). Changes in whole-brain functional networks and memory performance in aging. *Neurobiol Aging.* 35, 2193-2202.

Sandell, J.H., Peters, A., (2003). Disrupted myelin and axon loss in the anterior commeasure of the aged rhesus monkey. *J Comp Neurol.* 466,14–30.

Sani, S., Traul, D., Klink, A., Niaraki, N., Gonzalo-Ruiz, A., Wu, C.K., Geula, C., (2003). Distribution, progression and chemical composition of cortical amyloid-beta deposits in aged Rhesus monkeys: similarities to the human. *Acta Neuropathol (Berl).* 105, 145-56.

Sims-Robinson, C., Hur, J., Hayes, J.M., Dauch, J.R., Keller, P.J., Brooks, S.V., Feldman, E.L., (2013). The role of oxidative stress in nervous system aging. *PLoS One.* 8, e68011.

Smith, K.J., Kappor, R., Felts, P.A., (1999). Demyelination: The role of reactive oxygen and nitrogen Species. *Brain Pathol.* 9, 69-92.

Solís-Chávez, S.A., Durand-Rivera, A., Ibáñez-Contreras, A., Reyes-Pantoja, S.A., Valderrama, K., Heras-Romero, Y., Tena-Betancourt, E., Galván-Montaño, A., Alfaro-Rodríguez, A., Hernández-Godínez, B., (2014). Visual Evoked Potentials to Light Flashes in Captive Rhesus Monkeys: A Study Reflecting Cerebral Cortical Activity and Brain Maturation. *Pak Vet J.* 34, 41-45.

Tabbal, S.D., Tian, L., Karimi, M., Brown, C.A., Loftin, S.K., Perlmutter, J.S., (2012). Low nigrostriatal reserve for motor Parkinsonism in nonhuman primates. *Exp Neurol.* 237, 355-362.

Tigno, X., Gereanich, G., Hansen, B., (2004). Age-related changes in metabolic parameters of Nonhuman primates. *J. Gerontol. A. Biol. Sci. Med. Sci.* 59, 1081-1088.

Toledano, A., Álvarez, M.I., López-Rodríguez, A.B., Toledano-Díaz, A., Fernández-Verdecia, C.I., (2012). Does Alzheimer's disease exist in all primates? Alzheimer pathology in non-human primates and its pathophysiological implications (I). *Neurologia.* 27, 354-369.

Torre, P. 3rd, Mattison, J.A, Fowler, C.G, Lane, M.A, Roth, G.S, Ingram, D.K., (2004). Assessment of auditory function in rhesus monkeys (Macaca mulatta): effects of age and calorie restriction. *Neurobiol Aging* 25, 945-954.

Torre, P. 3rd, Fowler, C.G., (2000). Age- related changes in auditory function of rhesus monkeys (Macaca mulatta). *Hear Res* 142, 131-140.

Torre, P. 3rd, Lasky, R.E, Fowler, C.G., (2001). Aging and middle ear function in Rhesus monkeys (Macaca mulatta). *Audiology* 39, 300-304.

Tumminia, S.J., Chambers, C., Qin, C., Zigler, J.M. Jr., Russell, P., (1996). A comparison of antioxidant enzyme activities in organ-cultured rhesus monkey lenses following peroxide challenge. *Curr Eye Res.* 15, 845-851.

Valenzuela, R., Barroso-Chinea, P., Villar-Cheda, B., Joglar, B., Muñoz, A., Lanciego, J.L., Labandeira-Garcia, J.L., (2010). Location of protein receptors in primate substantial nigra: effects on dopaminergic cell death. *J Neuropathol Exp Neurol.* 11, 1130-1142.

Von Bernhardi, R.M., (2005). Envejecimiento: cambios bioquímicos y funcionales del sistema nerviosa central. *Rev Chil Neuro-Psiquiat*. 43, 297-304.

Wang, M., Gamo, N.J., Yang, Y., Jin, L.E., Wang, X.J., Laubach, M., Mazer, J.A., Lee, D., Arnsten, A.F., (2011). Neuronal basis of age-related working memory decline. *Nature*. 476, 210-213.

Watson, S.N, Nelson, M.A, Wildering, W.C., (2012). Redox agents modulate neuronal activity and reproduce physiological aspects of neuronal aging. *Neurobiol Aging*. 33, 149-161.

Willette, A.A., Gallagher, C., Bendlin, B.B., McLaren, D.G., Kastman, E.K., Canu, E., Kosmatka, K.J., Field, A.S., Alexander, A.L., Colman, R.J., Voytko, M.L., Weindruch, R.H., Coe C.L., Johnson, S.C., (2010). Homocysteine, neural atrophy, and the effect of caloric restriction in Rhesus monkeys. *Neurobiol Aging*.33, 670-680.

Yue, F., Zeng, S., Wu, D., Yi, D., Alex Zhang, Y., Chan P., (2012). Age-related decline in motor behavior and striatal dopamine transporter in cynomolgus monkeys. *J Neural Transm.* 119, 943-952.

Biographical Sketch

Ibáñez Contreras Alejandra

Education:

2012–2017	Ph.D. student in Biological sciences and health. Department of Health Sciences, Universidad Autónoma Metropolitana, Iztapalapa. México D.F, P.C. 09340.
2003–2008	Doctor of Veterinary Medicine (D.V.M.), Universidad Autónoma Metropolitana, Xochimilco. México D.F, P.C. 04960.
2005–2007	Social Service, in the Laboratory of Neuroprotection, Instituto Nacional de Rehabilitación (INR),

Languages: Spanish and English

Professional Experience:

2013–present	Medical Director in Applied Research in Experimental Biomedicine S.A. de C.V. (APREXBIO).
2008–present	Doctor of Veterinary Medicine (D.V.M.), assigned to the bioterio of BIOINVERT (Biología Integral para Vertebrados), Animal Experimental Unit.
2005–2016	Doctor of Veterinary Medicine (D.V.M.), assigned in non human primates unit, Centro de Investigación Proyecto Camina A.C.

Recognition and Awards

- Second Prize: 1st Congress of Students of PhD in Biological and Health Sciences as an author: Relación del estrés oxidante sobre la actividad eléctrica cerebral de las vías sensoriales (auditiva, visual y somatosensorial) en monos rhesus geriátricos (Macaca mulatta). Awarded for the Doctorado en Ciencias Biológicas y de la Salud, Universidad Autónoma Metropolitana, 25 and 26 of September 2014.
- First Prize "Académico Dr. Manuel J. Catillejos" as coautor of the work "Cirugía intrauterina: Proyecto de investigación modelo en primates no humanos *Macaca mulatta.*" Awarded for the Academia Mexicana de Cirugía, México D.F. 24 November 2009.
- First Prize "Académico Dr. Francisco Fonseca García" as coautor of the work "Desarrollo de un modelo animal de mielomeningocele y opciones de tratamiento prenatal en *Macaca mulatta.*" Awarded for the Academia Mexicana de Cirugía, México D.F. 25 November 2008.
- Third place in the poster exhibition, VII Annual Research Meeting, entitled "Establecimiento de los valores normales de las latencias en Potenciales Provocados Visuales por medio de flash en un grupo de monos rhesus (*Macaca mulatta*) en cautiverio." Awarded for the Instituto Nacional de Rehabilitación. México D.F. 21 - 23 de November 2007.
- First National Surgical Award "Académico Dr. Francisco Montes de Oca" como coautor del trabajo "Comparación de dos técnicas quirúrgicas para la reparación intrauterina del mielomengocele en *Macaca mulatta.*" Awarded for the Secretaria de Salud y la Academia Mexicana de Cirugía, México D.F. 28 November 2006.

Journalist Note

- Cristerna Salvador. Reparan malformación con cirugía intrauterina. CON CIENCIA, El Universal. 23 April 2007. P. F4.

Research Projects:

- Evaluación de la Seguridad y Eficacia del Y02-AK. Research Associate: Ibáñez Contreras Alejandra, in the period September 2012 to December 2012.
- Evaluación del suplemento alimenticio SM para diabéticos en Macaco rhesus (*Macaca mulatta)* con diabetes espontánea. Research Associate: Alejandra Ibáñez Contreras. In In the period December 2010 to April 2011.
- Evaluación de la eficiencia y seguridad preclínica en primates no humanos de una vacuna recombinante contra el virus de la hepatitis E (PROB-E). Research Associate: Ibáñez Contreras Alejandra, in the period October 2008 to June 2009.
- Bio-compatibility tests in bone and soft tissues using experimental hemostatic agents, dental restorative materials and endodontic sealers. Research Associate: Alejandra Ibáñez Contreras. In the period September 2007 to November 2011.
- Desarrollo de un modelo sencillo, graduable y reproducible de lesión traumática de médula espinal en primates. Research Associate: Alejandra Ibáñez Contreras. In the period October 2005 to November 2007.
- Comparación de dos técnicas quirúrgicas para la reparación intrauterina del mielomeningocele en *Macaca mulatta.*" Research Associate: Alejandra Ibáñez Contreras. In the period November 2005 to September 2007.

Publications with Imapact Factor:

- Visual, auditive and somatosensory pathways alterations in geriatric rhesus monkeys (*Macaca mulatta).*
  Alejandra Ibáñez-Contreras, Adrian Poblano, Marcela Arteaga-Silva, Braulio Hernández-Godínez, Ulalume Hernández-Arciga, Rafael Toledo, Mina Königsberg. In: Journal of Medical Primatology (J Med Primatol) In press 2016. DOI: 10.1111/jmp.12211. ISSN: 1600-0684 (online)
- Spinal Cord Injury of Rhesus Monkey Implanted with PPy/I Plasma Polymer, MRI Study.
  Axayacalt Morales-Guadarrama, Hermelinda Salgado-Ceballos, Israel Grijalva, Juan Manuel Camilo Ríos, Guillermo J Cruz, Araceli Díaz-Ruíz, María Guadalupe Olayo, Laura Álvarez Mejía, Rodrigo Mondragón Lozano, Alejandra Ibáñez Contreras, Braulio Hernández Godínez, Roberto Olayo. VI Latin America Congress on Biomedical Engineering CLAIB 2014, Paraná, Argentina 29, 30 and 31 October 2014, IFMBE Proceedings

49, DOI: 10.1007/978-3-319-13117-7_46. Braidot A and Hadad A. (eds). Springer International Publishing Switzerland 2015.

- Visual Evoked Potentials to light flashes in captive rhesus monkeys: A study reflecting cerebral cortical activity and brain maturation.
  Salvador A Solís-Chávez, Alfredo Durand-Rivera, Alejandra Ibáñez-Contreras, Sergio A. Reyes-Pantoja, Karla Valderrama, Yessica Heras-Romero, Eduardo Tena-Betancourt, Alfonso Galván-Montaño, Alfonso Alfaro-Rodríguez, Braulio Hernández-Godínez. In Pakistany Veterinary Journal (Pak Vet J). 2014, 34(1): 41-45.
- Absence of the right wing of the ilium generated a compensatory scolisos in rhesus monkey in captivity: diagnostic imaging and its alteration on the SEPTN.
  Salvador A. Solís-Chávez, Alejandra Ibáñez-Contreras, Sergio A Reyes-Pantoja, Gerardo Perdigón-Castañeda, Braulio Hernández-Godínez. In Journal of Medical Primatology (J Med Primatol). 2014 Feb;43(1):44-7.
- Multiple extrauterine pregnancy with early and near full-term mummified fetuses in a New Zealand white rabbit (*Oryctolagus cuniculus*).
  Eduardo Tena-Betancourt, Carlos A Tena-Betancourt, Alejandra M Zúniga-Muñoz, Braulio Hernández-Godínez, Alejandra Ibáñez-Contreras, Verónica Graullera-Rivera. In Journal of the American Association for Laboratory Animal Science (JAALAS). 2014 Mar;53(2):204-7.
- Changes in blood parameters in rhesus monkeys (Macaca mulatta) during the first trimester of gestation.
  Alejandra Ibáñez-Contreras, Braulio Hernández-Godínez, Sergio A. Reyes-Pantoja, Andrea Jiménez-García, Salvador A. Solís- Chavéz, Rodrigo Suarez-Gutiérrez, Alfonso Galván-Montaño. In Journal of Medical Primatology (J Med Primatol). 2013; 42: 171-176.
- Morphology of Brain Waveforms and Electrical Activity of Visual Evoked Potentials (VEP) of the Iguana iguana Species under Captivity Conditions in Mexico.
  Salvador Abraham Solís-Chávez, Alejandra Ibáñez-Contreras, Alfredo Durand-Rivera, Sergio Reyes-Pantoja, Alfonso Alfaro-Rodríguez, Eduardo Tena-Betancourt, Braulio Hernández-Godínez. In Thai Journal of Veterinary Medicine (Thai J Vet Med). 2013. 43: 113-117.
- DTI Tractography and MRI in chronic cerebral ischemia in Rhesus Monkey (Macaca mulatta) in vivo.
  Axayacatl Morales-Guadarrama, Iván Mejia-González Braulio Hernández-Godínez, Alejandra Ibáñez-Contreras, Sergio A. Reyes-Pantoja, Roberto Olayo. In Arquivos Neuropsiquiatria (Arq Neuropsiquiatr). 2013; 71: 134.

- Valores de latencia y respuestas electrofisiológicas de los diferentes componentes de los potenciales evocados auditivos del tallo cerebral en mono rhesus a diferentes intensidades (70-30 dB).
  Alejandra Ibáñez-Contreras, Braulio Hernández-Godínez, Alfredo Durand-Rivera. In Archivos de Medicina Veterinaria (Arch Med Vet) 2012; 44: 267-271.
- Preliminary Studies of Neurosensory and Cardiopulmonary Health Compromise in Captive– Bred Rhesus Macaques (Macaca mulatta) Suffering Scoliosis.
  Braulio Hernández-Godínez, Alejandra Ibáñez-Contreras, Sergio Reyes-Pantoja, Alfredo Durand-Rivera, Alfonso Galvan-Montaño, Gerardo Perdigón-Castañeda, Ma. Carmen Climent-Palmer, Eduardo Tena-Betancourt. In Journal of Medical Primatology (J Med Primatol) 2012; 41: 163-171.
- Piracetam-induced changes on the brainstem auditory response in anesthetized juvenil rhesus monkeys (Macaca mulatta). Report of two clinical cases.
  Alfredo Durand-Rivera, Rigoberto González-Piña, Braulio Hernández-Godínez, Alejandra Ibáñez-Contreras, Antonio Bueno-Nava, Alfonso Alfaro- Rodríguez. In Journal Medical Primatology (J Med Primatol). 2012; 41: 336-339.
- Clinical evaluation in geriatric rhesus monkeys (Macaca mulatta): through serological studies and physiological constants in captivity.
  Alejandra Ibáñez-Contreras, Braulio Hernández- Godínez, Sergio A. Reyes-Pantoja, Andrea Jiménez-García, Erika Morales-Cruz, Tonatiuh P Justo-Juarez, Alfonso Galván-Montaño. In Journal of Animal and Veterinary Advances (JAVA) 2011; 10: 3281-3286.
- New approach for graded compression spinal cord injuries in Rhesus macaque: method feasibility and preliminary observations
  Gabriel Guízar-Sahagún, Israel Grijalva, Braulio Hernández-Godínez, Rebecca E. Franco-Bourland, Leticia Cruz-Antonio, Angelina Martínez-Cruz, Alejandra Ibáñez-Contreras, Ignacio Madrazo. In Journal of Medical Primatology (J Med Primatol) 2011; 40: 401-413.
- Characterization of spontaneous subclavian steal phenomenon (SSP) in a female rhesus macaque (*Macaca mulatta*): A case report.
  Alejandra Ibáñez-Contreras, Braulio Hernández-Godínez, Gerardo Perdigón-Castañeda y Eduardo Tena-Betancourt. In Journal of the American Association for Laboratory Animal Science (JAALAS). 2011, 50; 1-3.

- Potenciales evocados auditivos del tallo cerebral en monos rhesus *(Macaca mulatta)* en diferentes etapas fisiológicas en condiciones de cautiverio.
  Alejandra Ibáñez-Contreras, Alfredo Durand-Rivera, Braulio Hernández-Godínez, y Sergio Abraham Reyes-Pantoja. Archivos de Medicina Veterinaria (Arch Med Vet), 2011 Vol. 43, 223-232.
- Valoración de latencias a través de los potenciales evocados somatosensoriales de nervio medial y tibial en monos rhesus, (*Macaca mulatta*) sometidos a laminectomía intrauterina para simular mielomeningocele.
  Braulio Hernández-Godínez, Alejandra Ibáñez-Contreras, Alfredo Durand-Rivera, Alfonso Galván-Montaño, Sergio Abraham Reyes-Pantoja, Eduardo Cárdenas-Lailson. Cirugía y Cirujanos (Cir Cir) 2011, 79: 237-241.
- Somatosensory evoked potentials SEPs, of median and tibial nerves in rhesus monkeys (*Macaca mulatta*) under captivity: Influence of ontogenic status in neonatal, infant, young, adult and senile stages.
  Braulio Hernández-Godínez, Alejandra Ibáñez-Contreras, Alfredo Durand-Rivera, Sergio A Reyes-Pantoja, Roberto Ramírez-Hernández, Paris Rodríguez-Guzmán y Eduardo Tena-Betancourt. In Journal of Medical Primatology (J Med Primatol) 2011 40:79-87.
- Radiographic findings of spine osteopatologies in rhesus monkeys *(Macaca mulatta*) in captive conditions.
  Braulio Hernández- Godínez, Alejandra Ibáñez-Contreras, Gerardo Perdigón-Castañeda, Alfonso Galván-Montaño, Guadalupe García-Montes de Oca, Carinthia Zapata-Valdez, Eduardo Tena-Betancourt. In Comparative Medicine (Comp Med) 2010; 60: 396-399.
- Hallazgos de anticuerpos contra Leptospira sp., serovariedades Panama, Lai, Australis, Shermani y Patoc, en un grupo de monos rhesus (*Macaca mulatta*) en condiciones de cautiverio.
  Alejandra Ibáñez-Contreras, Braulio Hernández-Godínez, Jorge.I. Torres-Barranca, y Patricia Meléndez-Velez. In Archivos de Medicina Veterinaria (Arch Med Vet). 2010. 42, 101-104.
- Anesthetic management in intrauterine surgery to evaluate an experimental model of myelomeningocele in non human primates (*Macaca mulatta*).
  Alfonso Galván-Montaño, Braulio Hernández-Godínez, Alejandra Ibáñez-Contreras, Luis Eduardo Cárdenas-Lailson, Roberto Ramírez-Hernández, and Jacqueline Aragón-Inclán. Acta Cirurgica Brasileira (Acta Cir Bras). 2010; 25: 294-297.

- Desarrollo de un modelo animal de mielomengocele y opciones de tratamiento prenatal en *Macaca mulatta.*
  Alfonso Galván Montaño, Eduardo Cárdenas-Lailson, Braulio Hernández Godínez, Alejandra Ibáñez Contreras, Andrea Martínez del Olmo, Jacqueline Aragón Inclán. Cirugía y Cirujanos (Cir Cir). 2007; 75; 357-362.

Book Publication:

Ecofisiología evolutiva de sistema nervioso de *Iguana iguana*. Salvador Abraham Solís-Chávez, Braulio Hernández-Godínez, Alejandra Ibáñez-Contreras. Editorial Académica Española. 2014. ISBN: 978-3-8473-6227-2

Chapters in International Books:

1. A review of rare reproductive dysfunctions: A complex Phenomenon of mammals and women.
   Capítulo 6. En: Ectopic Pregnancy: Diagnosis, Management and Complications.
   Eduardo Tena Betancourt, Braulio Hernández-Godínez, Alejandra Ibáñez-Contreras, Verónica Graullera-Rivera, Daniel Muñoz-Jiménez. Editorial Nova Science Publishers. ISBN: 978-1-63463-286-7. Editors: Mallory T. Church. 2014.
2. Brainstem's auditory evoked potentials and intervals values in rhesus monkey (*Macaca mulatta*) at different intensities (70-30 dB).
   Alejandra Ibáñez-Contreras, Salvador Abraham Solís-Chávez, Eduardo Tena-Betancourt, Braulio Hernández-Godínez.
   Capítulo 1. En Monkeys: Brain development, Social and hormonal mechanisms and Zoonotic diseases. Editorial Nova Science Publishers, Inc. Pp3-34. ISBN: 978-1-63117-851-1 Editors: Hugo A. Barrera Saldaña 2014.
3. The use of non-human primates as bio-models for aging brain studies: Interventonal role of oxidative stress in demyelization processes.
   Alejandra Ibáñez-Contreras, Adrian Poblano, Braulio Hernández-Godínez, Adriana Alarcon-Aguilar, Marcela Arteaga-Silva, Eduardo Tena-Betancourt, Mina Königsberg. Capítulo 2. En Monkeys: Brain development, Social and hormonal mechanisms and Zoonotic diseases. Editorial Nova Science Publishers, Inc. Pp. 35-50. ISBN: 978-1-63117-851-1 Editors: Hugo A. Barrera Saldaña 2014.
4. Ontogeny of somatosensory evoked potetials of median and tibial nerves in Rhesus monkeys (Macaca mulatta): influence of dissociative anesthetic mixtures under captivity conditions.

Braulio Hernández-Godínez, Herlinda Bonilla-Jaime, Adrian Poblano, Marcela Arteaga-Silva, Ma. Alejandra Bautista-Rodriguez, Salvador Solís-Chávez, Yessica Heras-Romero, Eduardo Tena-Betancourt, Alejandra Ibáñez-Contreras.
Capítulo 3. En Monkeys: Brain development, Social and hormonal mechanisms and Zoonotic diseases. Editorial Nova Science Publishers, Inc. Pp.51-78. ISBN: 978-1-63117-851-1Editors: Hugo A. Barrera Saldaña 2014.

Conference Presentations (Oral):

International Conferences	27
National Conferences	27
Poster Presentation	4
Invited Conferences	6

In: Evoked Potentials and Electrical Stimulation
Editor: Timothy Eisenhower
ISBN: 978-1-53611-075-3

Short Communication

VISUAL EVOKED POTENTIALS IN THE RHESUS MONKEY

Gamaliel Mendoza-Cuevas[1,2]**,**
Braulio Hernández-Godínez[1,2]
and Alejandra Ibáñez-Contreras[1,*]
[1]Applied Research in Experimental Biomedicine
(APREXBIO S.A. de C.V.)., Colonia Héroes de Padierna,
Tlalpan, Ciudad de México, México
[2]Centro nacional de Investigación en Imagenología e
Instrumentación Médica (CI3M). Universidad Autónoma Metropolitana,
Iztapalapa, Ciudad de México. México

ABSTRACT

Visual evoked potentials (VEPs) are a non-invasive technique used in the evaluation of the retino-geniculate-cortical response; such responses are derivative from the correct function of the structures belonging to the visual pathway, which are the retina, lateral geniculate body and visual cortex. VEPs, as being a current sensitive-based technique are also a suitable means for understanding pathologies developed from the injury or malformations throughout the pathway.

[*] Corresponding author email: ibanez.alejandra@hotmail.com

Keywords: visual evoked potentials, VEPs, retino-geniculate-pathway, rhesus monkey

INTRODUCTION

The study of the sensory systems is important from anatomical and physiological perspectives, as they are the afferent platforms from which the perception of the external world will be formed.

The visual pathways is one of the most studied sensory systems in the health sciences, as it is one of the fundamental systems from which humans depend, this is remarkable important because vision is the process by means of which we make a representation of images around our environment (Kolb, 2013).

Evaluation of these sensory systems with non-invasive techniques allows to relay on values of reference to distinguish whether a situation needs to be considered as pathological or normal, examples of such techniques are the magnetic resonance imaging (MRI) and the electroencephalogram (EEG) or its derivates such as Evoked Potentials (EPs). However, it is important to know the functioning of underlying structures that give rise to the typical wave form morphology of a VEP.

VISUAL PATHWAY

The information proceeding from the photons of light requires to be filtered, with the aim of getting to the retina and form the image to be processed in the visual cortex. Each structure of the eye helps with this goal. The contractions and relaxation of the ciliary muscles allow the iris to contract or expand depending on the external light conditions, the lens gets accommodated for the refraction of light altogether with the cornea, it is after this step that light reaches the retina at the level of the photoreceptors for the conversion of the stimulus into electrical potentials (Burns and Lamb, 2004; Schmolesky, 2007).

In the retina, the cells whose task is to convert the information of the photons into electrical currents are the photoreceptors, located in the rear layer of the eyeball (Kolb, 2003; Klintworth andCummings, 2007).

It is known that the retina contains photoreceptors such as cones that respond differentially to waves in the range of the blue, yellow and red colors

(~500, 600 and 700 nm, respectively), while the second type of cells, rods, respond better to changes in luminance (Burn and Lamb, 2004).

The absorption of photons by photoreceptor cells constitutes a response tc light in a fashion that has been described elsewhere. Briefly, it has been described that reaction to light is comprised by reactions of the rhodopsin (R), being excited (R*) and beginning to diffuse laterally so it hits several G-proteins, driving them to release its subunit alpha (Gα), which moves laterally and exchanges guanosine diphosphate for a molecule of guanosine triphosphate (Gα+GTP= G*), bounding to the phosphodiesterase (PDE), activating the α and β subunits, forming PDE**, catalyzing the hydrolysis of cGMP (cG) leading to the closure of nucleotide gated channels and blockage of the inward flux of Na^+ and Ca^{2+}, the Na^+/Ca^{2+}, K^+ exchanger sends Ca^{2+} out and the cytoplasmic Ca^{2+} declines during illumination (Burn and Lamb, 2004; Hurley, 2009).

The action potential created is transported throughout the retina layers by the on-and-off bipolar cells, regulated by the action of horizontal cells onto the ganglion cells, the latter form the optic nerve and carries the electrical pulse to the lateral geniculate body, taking part of both hemifields through the optic chiasma (Kolb, 2003; Creel, 2015).

The information of both hemifields arrives to the lateral geniculate body (LGB), which is a structure localized in the thalamus, composed by six layers that are divided in three pathways, carrying specific information of the retinal pattern, these pathways are known as the Magnocellular (M), Parvocellular (P) and Koniocellular (K), this structure projects information to the visual cortex trough the optic radiations (Campos-Ortega, Glees and Neuroff, 1968; Jeffries, Killian and Pezaris, 2014; Kaplan, 2004; Tobimatsu, Tomoda and Kato, 1995).

The M pathway has projections on the two more ventral layers of the LGB, transports information of fast temporal changes, codes the spatial properties of a stimulus, responding to achromatic stimulus (scotopic conditions), the P pathway has relay cells on the four more dorsal layers, proceeds from the photoreceptors that produce color vision and some cells that indicate changes in luminance (Kaplan, 2004, Tobimatsu, Tomoda and Kato, 1995). Finally, the K pathway, which is projected from the in-between layers of the M and P layers of the LGB, its cells generate potentials by the response of cones activated mostly by blue tonalities (Jeffries et al., 2014).

The segmented information arrives to the visual cortex, located at the posterior part of the brain, in the occipital lobe; it is in this zones where the information will be analyzed from its finest properties such as directions or movement (Hubel and Wiesel, 1968) until its identity (Grill-Spector and Malach, 2004).

At this level, the information arrives to the layer IV-C of what is the brodmann area 17 in humans, the M pathway arrives to the layer IV-Cα, while the P pathway gets to the layer IV-Cβ, some connections are formed between layer VI of V1, giving feedback to LGB, the K pathway arrives to layer IV-B (Kaplan,2004; Schmolesky, 2007), and from this layer the information is taken into different cortical areas (Van Essen, 2004), depending on what process is about to take place, for example the color of an object (V1, V2, V4 and V8), movement (V1, V2, MT, MTS, STS), identity of an object (V1, V3a, V4, V7, LOC), face recognition (VOT areas) and also places (PPA, Grill-Spector and Malach, 2004).

The 'visual circuitry' has been described in an explicit manner by Sherman (2001), in his work it is proposed that response of retina cells arrive through ionotropic receptors to the LGB, who sends the excitatory Post-synaptic potential (EPSP) to the layer IV of V1, in turn, layer VI sends modulatory signals to the thalamic reticular nucleus and interneurons of the LGB, operating between the retina and LGB, these potentials are generated in two ways, burst and tonically, being the former a 'wake-up call' that prepares the visual cortex for receiving more information of the visual environment.

VISUAL EVOKED POTENTIALS

Evaluating the activity aforementioned by means of non-invasive techniques is possible by using technology that allows that the organisms to stay alive, main point of the clinical approach of the neurosciences, and visual evoked potentials (VEPs) are a suitable model for this condition, this technique has the advantage of a better temporal resolution in comparison with techniques such as MRI.

Visual evoked potentials record the brain response to external stimulation, by averaging the bioelectrical response of the brain and displaying in an 'X' and 'Y' axis the neuronal response (Chiappa and Ropper, 1982).

VEPs have been developed as research techniques for evaluation of sensory pathways, also for the expression of the central nervous system (CNS) pathologies.

As such, VEPs are used for evaluating the functional integrity of the visual pathway through bioelectric signals produced by the CNS due to photic stimulation (Kothari, Bokariya, Singh and Singh, 2016), from the primary photoreceptor to the LGB in connection with the visual cortex (Creel, 2015;

Hofer and Frahm, 2008; Kothari, et al., 2016; Odom et al., 2016; Solís-Chávez et al., 2014; Ibáñez-Contreras et al., 2016).

One of the most used models of VEPs include photic stimulation of the entire visual field, besides of the pattern-VEP, the luminance VEP model proposes that there are cells in the retina that respond better to changes in luminance (Padmos et al., 1973; Van der Marel, Dagnelie and Spekreijse, 1984). Furthermore, this model is important as it is preferred in case subjects have poor visual acuity or are non-cooperative, like in the case of non-human primates (NHP).

In humans it has been widely accepted that the activity of the neural generators of the visual pathway activate at specific times, and the average of the electrical response throws a wave form that resembles a triangle, representing the activation of the LGB, V1 and association areas (Creel, 2015), and having an active (+) electrode in the occipital pole (Oz), the activity is seen at 75 ms, 100 ms and 145 ms (N1, P1 and N2 complex, respectively, Odom et al., 2016).

In NHP it is known they resemble human physiology, regarding same ontogenetic stages acquired in a shorter life-span. As to VEPs, Van der Marel and colleagues (Van der Marel et al., 1984) and Padmos (Padmos et al., 1973) have suggested that NHP show a similar response to visual stimulation, regarding on- and off-response of the visual cortex.

In our group, we tried to clarify the response of the rhesus monkeys to photic stimulations, since the values of such response have not been revealed in its totality. As can be seen in Table 1 and Figure 1, the response of a group of rhesus monkeys resemble the human response regarding the appearance of three deflections, achieved in our work by placing electrodes and set recording filters as the protocol used in humans (Jasper, 1958; Odom et al., 2016).

The values of the response in a sample of ten rhesus monkeys were about 45, 65 and 95 ms, which can be a proof for the suggestion of Woodman (2011) and Van der Marel and his group (Van der Marel et al., 1984) that NHP have a response 10-40 ms earlier than in human.

On the other side, there have been issues about the standardization of VEPs, since they are widely variable in the amplitudes they are recorded, these assumption represents a problem in the measurements as a departure point for the evaluation of the normal and pathological state, both, due to the noise introduced by the measured bio potentials and the technique (VEPs) itself.

Contributions to this field come from research into the form and materials of the electrodes used in the clinical neurophysiology.

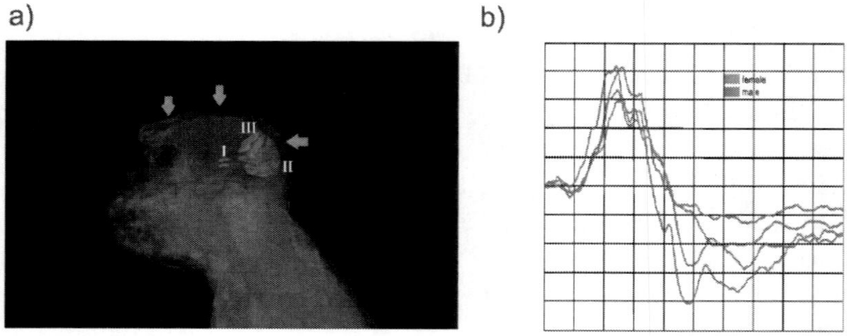

Figure 1. a) Generators of the electrical activity in the visual pathway, evaluated through VEPs and positioning of electrodes (arrows). I, lateral geniculate body (LGB). II, primary visual cortex (V1). III, association visual cortex. Arrows are indicative of the electrodes arrangement, being the near the occipitum the active electrode, in the frontal bone the reference electrode and the vertex was used as ground. b) Waveform of VEPs from a male and a female, recorded with cup electrodes.

It has been established that electrodes with different recording properties such as the covered area but different material can introduce electrical noise of up to 60 microvolts, besides the time of use is a variable that alters the voltage screening in the representation of electrical activity (Ferre, Clay and Tucker, 2001; Gondran et al., 1996); Huigen et al., 2002; Malmivuo and Suihko, 2004; Young et al., 2006).

This noise is dependent of the frequency at which the activity is registered, being less signal-to-noise ratio at low frequencies and the opposite is true for high frequencies (Huigen et al., 2002). However, these noise have been attributed to the amplifiers used in the measurement of VEPs.

Table 1. Values of latencies and amplitude differences of neural generators of the visual pathway in the rhesus monkeys (Macaca mulatta)

LATENCIES	**Cup electrode**	
Component	**Females**	**Males**
N1		
M(ms) ± DS	48.25(5.07)	46.38(2.92)
P1		
M(ms) ± DS	68.29(4.58)	68.88(5.96)
N2		
M(ms) ± DS	96.14(7.60)	94.31(7.29)

AMPLITUDES		
Components	**Females**	**Males**
N1-P1		
M(μV) ± DS	20.85(12.40)	12.28(3.92)
P1-N2		
M(μV) ± DS	22.01(11.24)	13.64(4.22)

Mean (M), standard deviation (SD) of latencies (expressed in milliseconds, ms) and amplitude differences of the components (in microvolts, μV) from the activity of the lateral geniculate body (LGB, N1), Primary visual cortex or V1 (P1) and extra striate visual cortex (N2)

Our work has given values for the VEPs in regard of amplitudes of acquisition, shown in Table 1. It is possible that VEPs in the rhesus monkey have several advantages accounting for its measurements as a reliable model for assessment of the visual pathway in a non-invasive manner.

By analyzing the difference of amplitude of the N1, P1 and N2 complex, the response across sexes seems to be similar, despite it is known that VEPs amplitudes are not readily to standardize due to individual differences (Chiappa and Ropper, 1982), therefore it is necessary to research the origin of the individual differences regarding VEPs amplitude, since it could be possible to even standardize the amplitudes of the brain response, arguing to Sherman's work (Sherman, 2001) in which both the response in tonic or burst mode does not change in amplitude, no matter what the strength of a previous neuronal response could be.

Hence, it is important to know to a more detailed extent, the response of an organism that is greatly important for biomedical research, since NHP and humans share physiological and morphological stages, being a model in which it is possible to prove drugs that aim to study its reactions in the brain to help preventing malformations of cells during elder age (Peters, 2009) or pathologies of the nervous system, and that can be achievable by having reference values from which any variable could be analyzed.

REFERENCES

Burns, M. and Lamb, T. Chapter 16: Visual transduction by rode and cone photoreceptors. En Chalupa, L. and Werner, J. (eds). (2004). *The visual neurosciences: volume 1.* Massachusetts: MIT Press.

Campos-Ortega, J.; Glees, P and Neuroff, V. (1968). Ultrastructural analysis of individual layers in the ñateral geniculate body of the monkey. *Zeitschrift für Zellforschung*, 87: 82-100.

Carlsson, H.; Schapiro, S; Farah, I. and Hau, J. (2004). Use of primates in research: a global overview. *Am J Primatol.* 63: 225-237. DOI: 10.1002/ajp.20054.

Chiappa, K. and Ropper, A. (1982). Evoked Potentials in clinical medicine (first of two parts). *The New England journal of medicine*; 306 (19): 1140-1150. Doi 10.1056/NEJM198205133061904.

Creel, D. (2015). Visually evoked potentials. *Webvision: The organization of the retina and visual system*; University of UTAH. Available from: http://webvision.med.utah.edu/

Elmore, L.; Ji, M.; Magnotti, J.; Leising, K.; Passaro, A.; Katz, J. and Wright, A. (2011). Visual Short-Term Memory Compared in Rhesus Monkeys and Humans. *Curr Biol. Jun*; 21 (11): 975-979. DOI: 10.1016/j.cub.2011. 04.031.

Ferree, T.; Clay, M. and Tucker, D. (2001). The spatial resolution of scalp EEG. *Neurocomputing*; Vol. 38-40: 1209-1216. Elsevier Press.

Gondran, C.; Siebert, E.; Yacoub, S and Novakov, E. (1996). Noise of surface biopotential electrodes based on NASICON ceramic and Ag-AgCl. *Medical and Biological Engineering and Computing*; 34: 460:466. doi:10.1007/ BF02523851.

Grill-Spector, K. and Malach, R. (2004). The human visual cortex. *Annu. Rev. Neurosci*. 27: 649-677.

Hofer, S. and Frahm, J. (2008). *In vivo* mapping of fiber pathways in the Rhesus Monkey Brain. *The open medical imaging journal*; 2: 32-41.

Hubel, D. and Wiesel, T. (1968). Receptive field and functional architecture of monkey striate cortex. *J Physiol.*, 195: 215-243.

Huigen, E.; Peper, A. and Grimbergen, C. (2002). Investigation into the origin of the noise of surface electrodes. *Medical and Biological Engineering and Computing*; 40: 332-338. doi:10.1007/BF02344216.

Hurley, J. (2009) Phototransduction. En Squire, L (ed.). Encyclopedia of Neurosciences. Elsevier.

Ibáñez-Contreras, A.; Poblano, A.; Arteaga-Silva, M. et al. (2016). Visual, auditive and somatosensory pathways alterations in geriatric rhesus monkeys (Macaca mulatta). *J Med Primatol,*; 45 (2): 92-102. DOI: 10.1111/ jmp.12211.

Jasper, H. (1958). Report of the committee on methods of clinical examination in electroencephalography: 1957. Electroencephalogr Clin Neurophysiol, 10 (2): 370-375.

Jeffries, A.; Killian, N and Pezaris, J. (2014). Mapping the primate lateral geniculate nucleus: A review of experiments and methods. *Journal of Physiology-Paris,* 108: 3-10.

Kaplan, E. Chapter 30: The M, P and K pathways of the visual system. E n Chalupa, L. and Werner, J. (2004) (eds). *The visual neurosciences: volume 1.* Massachusetts: MIT Press.

Kimotsuki, T.; Yasuda, M.; Tamahara, S.; Tomihari, M.; Matsuki, N. and Ono, K. (2006). Age-Associated Changes of flash visual evoked potentials in dogs. *J Vet Med Sci.*; 68(1): 79-82.

Klintworth, G. and Cummings T. (2007). *Capítulo 13: Normal eye and ocular adnexa.* En Mills, E. (Ed). Histology for pathologists (3rd Edition). Lippincott Williams and Wilkins.

Kolb, H. (2003). How the retina works. En Kolb, H.; Nelson, R.; Fernandez, E. y Jones, B. (s/f). *Webvision: The organization of the retina and visual system.* University of Utah. Recuperado de http://webvision.med.utah.edu/

Malmivuo, J. and Suihko, V. (2004). Effect of skull resistivity on the spatial resolution of EEG and MEG. IEEE Trans Biomed Eng; 51(7): 1276-1280. Doi 10.1109/TBME.2004.827255.

Odom, J.; Bach, M.; Brigell, M.; Holder, G.; McCulloch, D.; Mizota, A. and Tormene, A. (2016). *ISCEV standard for clinical visual evoked potentials: (2016 update);* Doc Ophtalmol. 133 (1): 1-9. DOI doi:10.1007/s10633-016-9553-y.

Padmos P, Haaijman J and Spekreijse H. (1973). Visually evoked cortical potentials to patterned stimuli in monkey and man. *Electroenceph Clin Neurophysiol.*; 35: 153–163. doi http://dx.doi.org/10.1016/0013-4694(73) 90171- 5.

Peters, A. (2009). The effects of normal aging on myelinated nerve fibers in monkey central nervous system. Frontiers in neuroanatomy. July 2009; 3 (Article11). DOI: https://doi.org/10.3389/neuro. 05.011.2009.

Solís-Chávez S, Durand-Rivera A, Ibáñez-Contreras A, Reyes-Pantoja S, Valderrama K, Heras-Romero, Y.; Tena-Betancourt E, Galván-Montaño A, Alfaro-Rodríguez A and Hernández-Godínez B. (2014). Visual evoked potentials to light flashes in captive rhesus monkeys: a study reflecting cerebral cortical activity and brain maturation. *Pak Vet J*; 34(1): 41-45.

Tu, H. and Hampton, R. (2014). Control of Working Memory in Rhesus Monkeys (Macaca mulatta). *J Exp Psychol Anim Learn Cogn.*; 40 (4): 467-476. Doi http://dx.doi.org/10.1037/xan0000030.

Van Der Marel E, Dagnelie G and Sprekreijse H. Subdurally recorded pattern and luminance EPs in the alert Rhesus Monkeys. Electroenceph Clin Neurophysiol. 1984; 57: 354-368. doi http://dx.doi.org/10.1016/0013-4694 (84)90159- 7.

Tobimatsu, S.; Tomoda, H. and Kato, M. (1995). Parvocellular and magnocellular contributions to visual evoked potentials in humans: stimulations with chromatic and achromatic gratings and apparent motion. *Journal of the Neurological Sciences,* 134: 73-82.

Van Essen D. (2004). Chapter 32: Organization of visual areas in Macaque and Human cerebral cortex. En Chalupa L and Werner J, editors. *The visual neurosciences: volume 1.* Massachusetts: MIT Press.

Young, G.; Ives, J.; Chapman, M. and Mirsattari, S. (2006). A comparison of subdermal wire electrodes with collodion-applied disk electrodes in long-term EEG recordings in ICU. *Clinical Neurophysiology*. 117 (6): 1376-1379. DOI: 10.1016/j.clinph. 2006.02.006.

Woodman, G. (2011). Homologues of Human ERP Components in Nonhuman Primates. In Luck, S. and Kappenman, E. *The Oxford Handbook of Event-Related Potential Components.* Oxford University Press.

INDEX

A

B

C

D

E

G

H

I

L

M

N

O

P

R

S

T

U

V